Outline of plant classification

Outline of
plant classification

Sandra Holmes

 LONGMAN   London & New York

Longman Group Limited
Longman House, Burnt Mill, Harlow
Essex CM20 2JE, England
Associated companies throughout the world

*Published in the United States of America
by Longman Inc., New York*

© Longman Group Limited 1983

First published 1983

British Library Cataloguing in Publication Data

Holmes, Sandra
   Outline of plant classification.
   1. Botany—Classification
   I. title
   581′ .012   QK95
   ISBN 0-582-44648-1

Library of Congress Cataloging in Publication Data

Holmes, Sandra, 1945—
   Outline of plant classification.

   Bibliography: p.
   Includes index.
   1. Botany—Classification.   I. Title.
QK97.H64   1983   581′ .012   82-13040
ISBN 0-582-44648-1

Set in 10/12pt Lasercomp Ehrhardt
Printed in Singapore by
Kyodo Shing Loong Printing Industries Pte Ltd.

Contents

Preface

In recent years plant classification has changed. There have been two causes – the introduction of standard endings (suffixes), and the incorporation of new data arising from the application of modern botanical techniques. The standard suffixes have resulted in categories acquiring unfamiliar new names; for example, algae are now known as Phycophyta, angiosperms as Magnolio-phytina. The incorporation of new data has resulted in groupings being changed; for example, in some schemes the ferns, gymnosperms and angiosperms collectively are classified as Pteropsida, while in other schemes the ferns alone are known as Pteropsida, or Filicopsida, or Filicatae, or Polypodiopsida.

The purpose of this book is to explain how and why names have been changed, and to describe the characteristics of each taxonomic group. It is intended to be helpful to the student trying to gain a groundwork in the plant kingdom, and to any more experienced or specialized botanist who needs to find his way through the unfamiliar framework of modern classification.

There is a good deal of terminology associated both with the taxonomy and the plant morphology used in the book. To cope with this there is an introduction to plant classificiation at the beginning and a glossary, mainly of plant morphology, at the end. The introduction is to enable the reader to remind himself of the order and status of taxonomic groups, and to become familiar with the terminology of the suffixes -phyta, -opsida, etc., before embarking on the application of these principles to the taxonomic groups. The glossary defines botanical words used in the text that may be unfamiliar to some readers. As most words occur more than once, it seems better to do this than to define them in the text at their first appearance, especially as some terms appear in several groups.

After the taxonomic introduction there is a section on the large-scale rearrangements of the plant groups, the higher taxa. The bulk of the book then takes each major group in turn, giving a summary of its classification, and describing the characteristics of each taxon. Since the aim of the text is to enable the reader to follow the different systems of classification rather than to choose a particular scheme, alternative systems of classification are described before or within the taxon to which they apply.

Classification is taken down to the order level in all groups. Except for angiosperms, examples are given of genera characteristic of each order. For the angiosperms, examples of families within each order are listed instead, since such families are well known to most botanists and provide a more useful indication than genera of the scope of the order.

The characteristics of groups of equal status are given in such a way that they can easily be compared, usually in the same kind of terms in the same sequence. Characteristics of the groups are in note form with numbers as figures, but comments and explanations are in normal sentences with numbers written out.

1 The classification of plants

The study of the grouping of plants together is known as systematics or taxonomy. These two terms are often used synonymously, but more accurately taxonomy is defined as the study of the principles and practice of classification, while systematics is the study of variation in living organisms. It is the aim of the taxonomist to produce a classification which best expresses the similarity and, it is hoped, the relationship between living organisms, and between them and their fossil relatives. This kind of classification is often called natural – meaning that it tries to express relationships, and that members of the same group are considered to have a common ancestor; in practice it is a classification based on numerous attributes.

The taxonomic groups

Plants are grouped into a taxonomic hierarchy. Each level of the hierarchy is called a **rank** or **category**, e.g. family, genus, species. A taxonomic group of any rank is called a **taxon** (plural **taxa**), e.g. Ranunculaceae, *Ranunculus*, *Ranunculus repens*.

The basic rank of the hierarchy is the **species**. A species is defined as: a group of plants which can interbreed together and form fertile offspring. Although this definition is unequivocal, it is not always practical. For example, it cannot be used on plants reproducing only non-sexually, on fossils, on dead or immature specimens collected in the field, or for plants with a long period before maturity. It is also not practical to carry out breeding experiments for every possible existing species. Besides this, some otherwise good species interbreed, e.g. many orchids. So a more practical definition of a species is: a group of plants which have a large number of characteristics in common, and are thought to have the capacity to breed together and to have originated from a common ancestor.

Species are grouped into **genera** (singular **genus**). A genus is a group of species thought to have a fairly recent common ancestor, but which do not interbreed together, or if they do so they form sterile hybrids. In plants, sterile hybrids can be made fertile by polyploidy, the doubling of the chromosome number. This has been done in the laboratory, and there is evidence that some of our flora originated in this way; indeed some species have many levels of polyploidy. The polyploid is a new species, separate from its parents since it cannot breed with them.

Genera are grouped into **families**. Some of these are very natural, such as the Cruciferae and Gramineae, while others are more mixed. It has recently become the practice to split large families into smaller ones. For example, from the large family Liliaceae the agaves and yuccas have been removed to the Agavaceae, and in some systems the onions and their close relatives are removed to the Alliaceae. The extent to which this should be done is a matter of opinion amongst taxonomists, some asserting that it makes a more natural classification showing ancestral relationships, while others suggest that it makes families proliferate to the extent that classifications become less manageable. Families were formerly known as 'natural orders'.

Families are grouped into **orders**. At this level degrees of relationship become much more difficult to establish, as the ancestor common to the group becomes more remote. It is not unusual for families to be transferred from one order to another, or for families to be grouped into different orders in different classificatory systems, as more information from palaeontology, biochemistry, or electron microscope studies becomes available.

Orders are grouped into **classes**. These have fewer characteristics in common, and are intended to reflect basic important differences between groups. With recent work from many fields, some classes have been split up. For example, in fungi the long-defined class Phycomycetes has now been divided into six classes, although the Ascomycetes and Basidiomycetes remain the same. The Phycomycetes is thought to be a more heterogeneous group whose members are considered not to be closely related now that they have been studied more carefully.

Classes are grouped into **divisions** or phyla (singular **phylum**). The status of these is under discussion. Some authorities still recognize the large old groups of Algae, Fungi, Bryophyta, Pteridophyta and Spermatophyta as divisions, but others consider that many of these groups are of higher than division level, and that classes in the old system should be raised to division status. This will be considered further in the next chapter.

The highest level of classification is that of the **kingdom**. There were formerly said to be two kingdoms, the plant and animal kingdoms, but then the flagellates, claimed as both Algae and Protozoa, blurred the distinction. Now a more fundamental difference is thought to be that between the prokaryotes and eukaryotes. This too will be discussed in detail in the next chapter.

There are other categories. Below each rank there can be a group using the prefix sub-, as in **subdivision, subclass, subfamily, subgenus**; between the family and the genus is the **tribe**, and between the genus and the species is the **section**. Below the species is the **subspecies, variety and form**, whose status tends to be variable.

The important ranks of the taxonomic hierarchy are:

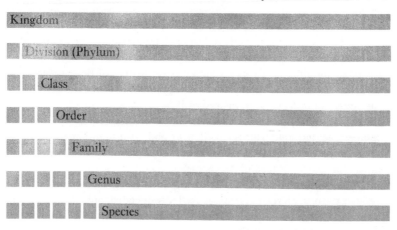

Kingdom

Division (Phylum)

Class

Order

Family

Genus

Species

The naming of taxa

The study of the naming of taxa is known as nomenclature, and the rules concerning it have been formalized in the International Code of Botanical Nomenclature (ICBN), which lays down conventions on the formation and use of scientific names. It requires that scientific names are in Latin form, and that they are subject to the rules of Latin grammar and written in the Latin alphabet, even if they are derived from other languages. The code is updated at regular intervals, and the information here is taken from the code adopted by the 12th International Congress in Leningrad in 1975 and published in 1978.

Names of taxa between kingdom and genus

The ICBN has stipulated standard endings for certain taxa, so that the rank of a taxon can always be recognized. The names below order level are standard for all plant groups. They are placed on the stem part of the name of a genus which is considered to be the type genus for the family. For example, the genus *Lilium* is a type genus; its stem is Lili- and the categories are made as follows:

Category	Ending	Example
Order	-ales	Liliales
Suborder	-ineae	Liliineae
Family	-aceae	Liliaceae
Subfamily	-oideae	Lilioideae
Tribe	-eae	Lilieae
Subtribe	-inae	Liliinae

Although these categories are mandatory there are eight non-standard exceptions, sanctioned by long usage, which are permitted as correct alternatives to family names ending in -aceae for eight families of flowering plants. The old alternative name reflects a characteristic of the group rather than the name of a type genus. The exceptions are:

Cruciferae, meaning petals arranged in a cross, instead of Brassicaceae.
Compositae, meaning compound flower head, instead of Asteraceae.
Gramineae, meaning grasses, instead of Poaceae.
Labiatae, meaning flowers having lips, instead of Lamiaceae.
Leguminosae, meaning fruit is a legume or pod, instead of Fabaceae.
Umbelliferae, meaning having flowers in umbels, instead of Apiaceae.
Palmae, the palms, instead of Arecaceae.
Guttiferae, meaning bearing droplets (of oil) instead of Clusiaceae.

Above the order level, the endings are recommended, not mandatory, and are different for fungi, algae and other non-fungal plants. The recommendation for the first part of the name is that it may be taken either from a distinctive characteristic of the group or from an included genus.

For non-fungal plants other than algae, the endings are:

Category	Ending
Division	-phyta
Subdivision	-phytina*
Class	-opsida*
Subclass	-idae
Superorder	-anae

*For subdivisions -icae and for classes -atae are still widely used, although they are not now recommended endings.

These endings can make long-familiar categories difficult to recognize. For example, *Magnolia* is the type genus of the Magnoliales and has been used as the type genus for the higher categories, the superorder Magnolianae, the class Magnoliopsida or Magnoliatae (the dicotyledons), and for the subdivision Magnoliophytina (the angiosperms), while the division is still known as Spermatophyta. These names are changed in classification systems which consider that angiosperms should be given division status; in this case the angiosperms are known as the division Magnoliophyta, but the dicotyledons still have class status and are still called Magnoliopsida.

The situation is made more confusing where an old name reflected a characteristic of the group rather than a type genus, and the new recommended ending is grafted on to either this old name or on to a type genus. For example, the horsetails, when considered a class of Pteridophyta, are known as the class Equisetatae (from the genus *Equisetum*) or as Articulatae (from Latin *articulus* = joint, i.e. jointed stems) or Sphenopsida (from Greek *sphen* = wedge, i.e. wedge-shaped leaves). The ferns are known as Filicopsida (from Greek *filix* = fern) or as Polypodiopsida (from the genus *Polypodium*). It is sometimes considered that the classes of pteridophytes should be raised to division status, in which case the ending -phyta replaces the class endings, but all the alternatives remain in use (see Ch. 8).

For algae the endings are:

Category	Ending
Division	-phyta
Subdivision	-phytina
Class	-phyceae
Subclass	-phycidae

This is fairly straightforward and usually poses no difficulties. It is more usual for the name to refer to a characteristic of the group rather than to a genus, such as Phaeophyta or Phaeophyceae for the division or class of brown algae (from Greek *phaeo* = brown). If the group Algae is considered to be of division status, it is renamed Phycophyta.

For fungi the endings are:

Category	Ending
Division	-mycota
Subdivision	-mycotina
Class	-mycetes
Subclass	-mycetidae

The division term -mycota is not widely used for fungi, although is seen occasionally as Mycomycota. More usually, as a division fungi are renamed as either Mycota or Mycophyta, using the higher plant division ending; or fungi are raised to the level of a kingdom and the two subgroups of fungi are given division status as Myxomycota and Eumycota (see Ch. 5).

Bacteria are covered by a separate code, the International Code of Nomenclature of Bacteria (ICNB). Here the endings are the same from the order down as for all other plants, and there are no standard endings above the order level.

Names of genera

The names of higher categories are written in Roman type, starting with a capital letter. The names of genera and species are conventionally written in italics in print and are underlined in manuscript. The generic name is a singular noun and begins with a capital letter, but can be any gender, e.g. *Lilium, Crocus, Amaryllis, Tagetes*.

Fossil finds cannot, of course, be subjected to breeding criteria to discover the range of structure within a genus, so the term form genus is more correctly used. Where fossils of stems, seeds, roots, etc. are found unconnected, even if together in the same rock formation, they must be assigned to different form genera. But if further work shows connections between them, a new name is given to the whole plant.

All names up to this rank are single words, known as uninomial, uninominal or unitary.

Names of species

The name of a species consists of two words, and is known as a binomial, binominal or binary. The first name is the generic name, and the second is the specific epithet, which is never used by itself. For example, in the genus *Ranunculus* (buttercups), *Ranunculus repens* and *Ranunculus aquatilis* are two species. Once the genus has been mentioned, it may be abbreviated to its first letter when subsequently writing the name of the species, as for example *R. repens* and *R. aquatilis*. It is more complete to place after the specific epithet the surname of the person who first named the species. The surname is written is a conventional abbreviated form, as in *Ranunculus repens* Linn. for Linnaeus. This is not done in general use, but is the correct form of the name and is seen in floras and in technical works.

Below the species level the terminology is complex and variable and is outside the scope of this book.

Anglicization of Latin names

It is customary to refer to a group formally with its correct Latin ending and using a capital letter, e.g. Bryophyta, Euglenophyta, but to anglicize these terms for more general use as e.g. bryophytes, euglenoids, changing the ending and reducing the first letter to lower case. So one would refer to the division Bryophyta or to bryophytes, depending on context. Although most names change both to a lower case and have a changed ending, algae and fungi and a few other groups do not. One refers to the Algae or the Fungi in a formal context and to algae and fungi in an informal one. In the text, formal and anglicized terms are used as the occasion demands; headwords are given in the correct latinized form, and the alternative names and anglicized form are given in parentheses (round brackets).

2 The organization of the higher taxa

It is at the level of the higher taxa that the most drastic reorganization of taxonomic groups has recently occurred, resulting in rearrangements of kingdoms, divisions and classes.

Prokaryotes and eukaryotes

As a result of the development of the electron microscope and chemical analysis techniques, it has become clear that bacteria and blue-green algae are related in a very fundamental way. The nuclear material (DNA) in this group is not arranged into chromosomes and there is no nuclear membrane, although the strand of DNA occupies a particular region of the cytoplasm called the nucleoid. For this reason, these two groups are placed together and

Table 2.1: Differences between prokaryotes and eukaryotes

Prokaryotes	Eukaryotes
Nuclear material not contained within nucleus; no chromosomes or nuclear membrane.	Nuclear material in nucleus and surrounded by nuclear membrane. Usually form chromosomes.
No mitochondria or plastids.	Mitochondria in most cells; plastids in plants and some flagellates.
Ribosomes small (70S type).	Ribosomes large (80S type).
Cell wall usually contains peptidoglycan (see Glossary).	This peptidoglycan absent.
Nuclear division not by mitosis or meiosis.	Nuclear division usually by mitosis or meiosis.
No sexual reproduction; but sexual processes in some bacteria.	Sexual reproduction common.
Flagellum, when present, does not have 9 + 2 structure.	Flagellum, when present has characteristic 9 + 2 structure, i.e. 9 outer fibrils and 2 inner ones.
Some can fix nitrogen	None can fix nitrogen.
Bacteria and blue-green algae.	All other plants and animals.

are known as prokaryotes (procaryotes). All other organisms, with a nucleus, nuclear membrane and other characteristics in common, are grouped together as eukaryotes (eucaryotes). Some major differences between the groups are given in Table 2.1.

The Prokaryota and Eukaryota may each be given the status of a kingdom.

Viruses are structurally simpler than either pro- or eukaryotes. They are not plants and are not considered here.

The Schizophyta

The prokaryotes are often considered to be a division known as the Schizophyta, with two classes, the Schizomycetes (bacteria) and Schizophyceae (blue-green algae). There has been some controversy as to how fundamental the division between pro- and eukaryotes may be, and blue-green algae are still included in works on algae, where they are ranked as a division called Cyanophyta or Myxophyta, or a class called Cyanophyceae or Myxophyceae.

Groups of eukaryotes

The Eukaryota may be organized in several ways. There may be considered to be three groups, namely the flagellates, and then the kingdoms of plants and animals which are thought to be descended from flagellate ancestors. The flagellates do not form a separate kingdom but are considered as algae or protozoans, and some of them as both.

Thallophyta and Embryophyta

The next level of taxonomic division of eukaryotes varies from one authority to another, but the terms Thallophyta and Embryophyta are sometimes used.

The Thallophyta comprises those forms in which the plant body is not made up of root, stem and leaves, but the composition of the group has varied. Originally it consisted of the algae, fungi, bacteria and viruses. This arrangement is now considered to be unsatisfactory and unhelpful, since it includes very heterogeneous groups, and the bryophytes are not included as thallophytes, but they too are not made up of true roots, stems and leaves. In some schemes, the Thallophyta is subdivided into the Protophyta, comprising all unicellular and loosely colonial plants, and the Thallophyta proper, comprising those plants whose body is a thallus, and including the bryophytes. Plants with true roots, stems and leaves are then known as Cormophyta (or Cormobionta), i.e. the Pteridophyta and Spermatophyta; the bryophytes are sometimes included in the Cormophyta because many have structures superficially resembling roots, stems and leaves.

The Embryophyta (or Embryobionta) comprises those forms in which a multicellular embryo develops from the zygote while the zygote is still attached to the parent. Gametangia and sporangia are multicellular and the outer layer is sterile and forms a wall, but in seed plants the gametangia are greatly reduced. Thus the Embryophyta comprises the Bryophyta, Pteridophyta and Spermatophyta.

It is sometimes considered that the Embryophyta is equivalent to the Archegoniatae, i.e. those plants possessing archegonia. A difficulty arises here that archegonia are present in a very reduced form in most gymnosperms and not at all in angiosperms, so the term Archegoniatae is usually used for the Bryophyta and Pteridophyta only.

The Embryophyta may be divided into the Bryophyta and Tracheophyta.

Bryophyta and Tracheophyta

The Bryophyta are embryophytes that do not possess vascular tissue; in all classifications they are considered to be a separate, discrete group.

The Tracheophyta are embryophytes that possess vascular tissue, taking their name from the tracheids or tracheary elements of the xylem. The sporophyte is always the dominant generation and the one in which vascular tissue develops. This group is believed to have originated from the algal group Chlorophyta, with earliest records in Silurian rocks. The Tracheophyta is considered to be a division (usually called a phylum), and is divided into four classes, the Psilopsida, Sphenopsida, Lycopsida and Pteropsida (although these classes may be raised to phylum status as Psilophyta, Sphenophyta, Lycophyta and Pterophyta). The Psilopsida are leafless primitive forms, sometimes known as whisk ferns, the Sphenopsida are the horsetails, and the Lycopsida are the club mosses; groups which in other classifications are included with the ferns in the Pteridophyta. In this scheme, the Pteropsida does not only include the ferns, but also the gymnosperms and angiosperms, on the grounds that they have basically large leaves with a similar internal anatomy and presumably a common ancestor because fossils suggest that spermatophytes arose from ferns.

Alternatively, the Psilopsida, Sphenopsida, Lycopsida and ferns together comprise the Pteridophyta, while the angiosperms and gymnosperms comprise the Spermatophyta. On this system the ferns alone may be known as Pteropsida, or may be called Filicopsida, Filicatae or Polypodiopsida.

The system used here is to consider the divisions to be Pteridophyta and Spermatophyta, the Pteridophyta comprising the classes Psilopsida, Sphenopsida, Lycopsida and Filicopsida, and the Spermatophyta comprising the subdivisions angiosperms and gymnosperms.

Angiosperms and gymnosperms

The status of angiosperms and gymnosperms is also under discussion as more details of their fossil history come to light. They were considered to be equal subdivisions of the division Spermatophyta, based on the fact that all are seed plants (spermatophytes), while the gymnosperms have naked seeds and the angiosperms have seeds enclosed in a fruit; other important differences between the two groups are discussed in Chapter 9.

Recently the status of the groups has been disputed, and it is now thought that the two groups of gymnosperms with a fossil history, the conifers and cycads, each had a distinct origin, as distinct as that separating them from the angiosperms, and that the seed habit may have originated independently in each group. So it is proposed that there should be three subdivisions of equal rank, which are known as the Coniferophytina (or Pinicae), the Cycadophytina (Cycadicae) and the Magnoliophytina (Angiospermae).

Phanerogamia and Cryptogamia

These are two obsolete terms which still find a use as collective common names. The Phanerogamia (phanerogams) are those plants with seeds, i.e. the Spermatophyta, while the Cryptogamia (cryptogams) are the rest, the plants without seeds; the term Cryptogamia is still used mainly to comprise the algae, fungi, lichens, bryophytes and pteridophytes.

The system used here

The chapters on the groups are arranged as shown on page 11.

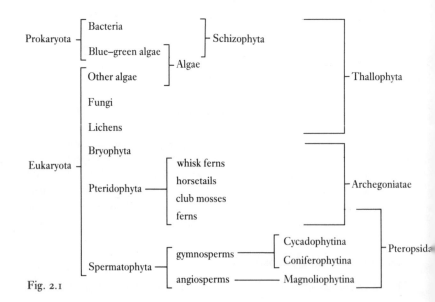

Fig. 2.1

Schizophyta
Algae
Fungi
Lichens
Bryophyta
Pteridophyta
Spermatophyta: gymnosperms
Spermatophyta: angiosperms

It is not suggested that these are necessarily the most natural groupings, but they are familiar and easy to handle. The relationship of these groups to the terminology discussed in this chapter is summarized in Figure 2.1.

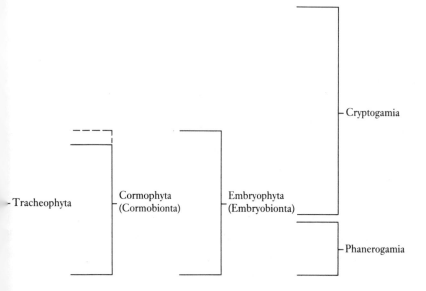

3 Schizophyta

At present, Schizophyta is the name most usually applied to the division containing prokaryotic organisms, but classification at this level is very fluid and variable. The scheme explained below endeavours to include all the terms the reader is likely to meet, and is not primarily the latest thinking on the subject. On evolutionary grounds the blue-green algae probably should be considered here rather than with the algae; but details of the group and its orders are given in the algal chapter, because nomenclature amongst the algae is more stable and established than amongst prokaryotes; but an outline of the blue-green algae is given here so that they can be compared with bacteria.

The authority on classification of bacteria is *Bergey's Manual of Determinative Bacteriology*. It is the system in the 7th edition of 1957, describing ten orders of bacteria, that has been most widely used in the last 20 years, but it is not satisfactory in that the nature of prokaryotes was not widely accepted at that time, so viruses were included with the rickettsias as orders of the class Microtatobiotes. The most recent, 8th edition of Bergey was published in 1974; it recognizes the separateness of prokaryotes and excludes viruses, but it reorganizes the prokaryotes in a very unfamiliar way. It does not assign all genera of bacteria to orders, on the grounds that there is, as yet, insufficient information to do so, but places the genera in 19 groups (or parts), some of which contain orders while others do not. The system described below is a modification of Bergey's 7th edition, maintaining the familiar ten orders of bacteria, but relating them to the 19 parts of the 8th edition of 1974. After this, there is a brief summary of the organization of the 8th edition, followed by Bergey's propositions for a further rearrangement in the future.

Summary of classification of prokaryotes

Division Schizophyta (Prokaryota, fission plants)

Class Schizophyceae (blue-green algae)

for orders and details, see Algae

Class Schizomycetes (bacteria, fission fungi)

Order 1 Pseudomonadales

Order 2 Chlamydobacteriales

Order 3 Hyphomicrobiales

Order 4 Eubacteriales

Order 5 Actinomycetales

Order 6 Caryophanales

Order 7 Beggiatoales

Order 8 Myxobacterales

Order 9 Spirochaetales

[Class Mollicutes] not always recognized

Order 10 Mycoplasmatales

[Class Microtatobiotes] no longer exists

Order Rickettsiales $\begin{bmatrix} \text{Rickettsiales} \\ \text{Chlamydiales} \end{bmatrix}$

[Order Virales] formerly, before recognized that they were not prokaryotes

The numbers 1 to 10 in the above represent the original ten orders of bacteria. The mycoplasmas are now recognized as fundamentally different from all other bacteria, so they are placed in a separate class, Mollicutes, whose characteristics are given below. Rickettsias were never considered to be one of the orders of true bacteria.

Division Schizophyta

Possess characteristics of prokaryotes: see Chapter 2.

Class Schizophyceae

(blue-green algae)

This class is described under algae as the division Cyanophyta. The characteristics given there are such that it can be compared with the other divisions of algae. Class characteristics are given here as well, so that it can be compared with bacteria, but it is stressed that this is the same group, with the same characteristics, as the division Cyanophyta:

Predominantly autotrophic. Photosynthesis leads to production of oxygen as in other green plants, and the hydrogen source is water. Chlorophyll present on thylakoids. Flagella absent, but some move by creeping or gliding. Many possess gas vacuoles. Food reserves include cyanophycean starch. No sexual reproduction, but some reports of genetic recombination; meiosis never observed. Most photosynthetic, but can lose chlorophyll, e.g. *Beggiatoa*, which is also classified as a bacterium. Photosynthetic pigments are chlorophyll *a*, carotenoids especially β-carotene, and the biloproteins phycocyanin and phycoerythrin. Form endospores and exospores.

For details and orders, see Cyanophyta.

Class Schizomycetes

(bacteria, fission fungi)

Includes both heterotrophic and autotrophic forms. Those which photosynthesise never do so with production of oxygen, and the hydrogen source is never water. Chlorophyll, when present, not on thylakoids. Flagella often present. Some possess gas vacuoles. Food reserves include starch-like and glycogen-like polysaccharides, fats and waxes. Sexual processes occur; meiosis never observed. Most saprophytic or parasitic, but obligate parasitism rare and parasites are extracellular (similar but smaller intracellular parasites being placed in the order Rickettsiales). Photosynthetic pigments include bacteriochlorophylls; do not possess biloproteins. Spores produced are endospores or occasionally exospores.

Order Pseudomonadales

Cells have rigid walls. Unicellular, usually rod-shaped, coccoid or spiral, occasionally forming chains; sometimes contain purple or green pigments. Usually motile by 1 or more flagella. Photosynthetic, chemosynthetic and parasitic. Never form spores; reproduction by fission.

e.g. *Acetobacter, Chlorobium, Chromatium, Desulfovibrio, Methanobacterium, Nitrobacter, Nitrosomonas, Pseudomonas, Rhodospirillum, Spirillum, Thiobacillus, Vibrio.*

In the 1974 system this order does not exist and its members are scattered through many groups. *Chlorobium, Chromatium* and *Rhodospirillum*, the pigmented photosynthetic bacteria, are placed in the order Rhodospirillales (part 1, p. 18), *Spirillum* is in part 6, *Pseudomonas* and *Acetobacter* in part 7, *Vibrio* is in part 8, *Desulfovibrio* in part 9, *Nitrobacter, Nitrosomonas* and *Thiobacillus* in part 12, and *Methanobacterium* in part 13.

Order Chlamydobacteriales

Cells have rigid walls. Form many-celled filaments (trichomes), frequently ensheathed in mucilage sometimes containing ferric hydroxide. Trichomes often attached to surface, but single cells may emerge either as non-motile spores or as zoospores with flagella. Saprophytic. Reproduction by spores or by fragmentation of filaments. Also called alga-like bacteria.

e.g. *Crenothrix, Leptothrix, Sphaerotilus.*

In the 1974 system this order does not exist but its members are in the group Sheathed Bacteria (part 3, p. 18).

Order Hyphomicrobiales

Cells have rigid walls. Unicellular or filamentous. Usually non-motile, often attached to surface by stalks. Saprophytic or photosynthetic. Reproduction by budding (as yeasts) rather than by fission, so also called budding bacteria. e.g. *Hyphomicrobium, Rhodomicrobium.*

In the 1974 system this order does not exist and its members are separated into different groups. *Hyphomicrobium* is a member of the Budding Bacteria (part 4) and *Rhodomicrobium* is a member of the Rhodospirillales (part 1).

▮▮▮ Order Eubacteriales

Cells have rigid walls. Unicellular, coccoid or rods, sometimes in chains or packets, but do not form trichomes. Motile by flagella, and non-motile forms. Parasites, saprophytes or symbionts. Reproduction by fission; spores formed in some species. Sometimes called true bacteria.
e.g. *Azotobacter, Bacillus, Bacteroides, Clostridium, Corynebacterium, Escherichia, Eubacterium, Lactobacillus, Methanococcus, Neisseria, Pasteurella, Rhizobium, Salmonella, Staphylococcus, Streptococcus, Veillonella.*

In the 1974 system this order does not exist, and its members are scattered amongst other groups, e.g. *Rhizobium* and *Azotobacter* in part 7, *Escherichia, Pasteurella* and *Salmonella* in part 8, *Bacteroides* in part 9, *Neisseria* in part 10, *Veillonella* in part 11, *Methanococcus* in part 13, *Staphylococcus* and *Streptococcus* in part 14, *Bacillus* and *Clostridium* in part 15, *Lactobacillus* in part 16, *Eubacterium* and *Corynebacterium* in part 17.

▮▮▮ Order Actinomycetales

(actinomycetes, ray fungi)

Cells have rigid walls. Cells rod-shaped but may grow out in a branching system resembling mould colonies, and produce conidia-like spores. Non-motile. Parasites, saprophytes and symbionts. Reproduction by fission or by spores. Many produce antibiotics.
e.g. *Actinomyces, Mycobacterium, Streptomyces.*

In the 1974 system this order still exists and is included in part 17.

▮▮▮ Order Caryophanales

Cells have rigid walls and are larger than in other orders. Cells in trichomes. Sometimes motile by lateral flagella. Saprophytic in water and decomposing matter and parasitic in intestines of arthropods and vertebrates. Reproduction by fission.
e.g. *Caryophanon, Simonsiella.*

In the 1974 system this order does not exist and its members are found in different groups, e.g. *Simonsiella* is in the Cytophagales (part 2) and *Caryophanon* is in part 16.

Order Beggiatoales

Cells have rigid walls. Cells coccoid and often in trichomes; do not possess photosynthetic pigments but otherwise resemble blue-green algae. Move by gliding movements, as blue-green algae; no flagella. Chemosynthetic, oxidize hydrogen sulphide to sulphur which they store in cells as granules. Reproduction by break-up of trichomes.

e.g. *Beggiatoa, Thiothrix*.

In the 1974 system this order does not exist and its members are found in the Cytophagales (part 2).

Order Myxobacterales

(Myxobacteriales, myxobacteria, slime bacteria)

Cells have flexible walls. Form a swarm (pseudoplasmodium) of small flexible coloured rods. Mostly saprophytic, but in artificial cultures some feed by lysis on other living organisms, such as other bacteria. Move by creeping on surfaces. Do not form spores, but can form accumulations held together by mucilage where rods converted to resting cells (cysts); reproduction by stalked fruit bodies which develop from spreading colonies like slime moulds.

e.g. *Archangium, Cytophaga, Myxococcus, Sporocytophaga*.

In the 1974 system this order is divided into two, the Myxobacterales and Cytophagales (both in part 2). The differences are:

Myxobacterales	Cytophagales
Unicellular, embedded in slime.	Unicellular or filamentous.
Reproduction by binary transverse fission.	Reproduction by hormogonia and gonidia.
Form fruit bodies of slime and cells.	Do not form fruit bodies.
e.g. *Myxococcus, Archangium*.	e.g. *Cytophaga, Sporocytophaga*, and members of the Beggiatoales.

Order Spirochaetales

Cells spiral and have flexible walls. Single-celled with very long slender body. Swim by flexion with snake-like movements. Free-living, saprophytes, parasites, or commensals. Reproduction by fission.

e.g. *Borrelia, Leptospira, Spirochaeta, Treponema*.

In the 1974 system the order exists in this form (part 5).

Class Mollicutes

Now held to be distinct from bacteria because:
Ultramicroscopic and can pass through bacterial filters. Cell wall and reproduction unlike that in bacteria. 1 order, Mycoplasmatales, whose characteristics are described below in the same way as the other nine orders, as this was the 10th order of bacteria.

Order Mycoplasmatales

(mycoplasmas)

Flexible cells, bounded by a 3-layered membrane, not a true cell wall, and cannot secrete cell wall precursors like muramic acid; very small. Highly variable in shape (pleomorphic), coccoid to filamentous. Usually non-motile. Saprophytic and parasitic. Reproduction controversial, but there seem to develop within filaments tiny coccoid structures called elementary bodies, released by fragmentation of filaments and/or binary fission; budding also occurs.
e.g. *Acholeplasma, Mycoplasma, Spiroplasma, Thermoplasma.*

Mycoplasmas are the cause of pleuropneumonia in cattle, so the Mycoplasmatales were formerly called pleuropneumonia-like organisms (PPLOs) sometimes given order status as Pleuropneumoniales or Borellomycetales. Mycoplasmas were thought only to exist in animals; when similar organisms were found causing diseases in plants, they were called mycoplasma-like organisms (MLOs) or mycoplasma-like bodies (MLBs). They cause disease of the yellows type, and were formerly considered to be viruses, but no virus was found, and organisms were discovered in the phloem resembling mycoplasmas. It may be that there should be a new order of Mollicutes to contain the plant parasites.

In the 1974 system the mycoplasmas are included in part 19.

Class Microtatobiotes

This is a class comprising the smallest living things, all of which are parasitic and most intracellular. This class has now disappeared because rickettsias are prokaryotes and viruses are not.

Order Rickettsiales

(rickettsias)

Small, rod-like, spherical or irregular intracellular parasites of man and animals. Similar organisms found in plants are called rickettsia-like organisms (RLOs). Possess typical bacterial-type cell walls and no flagella.
e.g. *Chlamydia, Coxiella, Ehrlichia, Rickettsia.*

In the 1974 system the rickettsias form part 18, and are divided into two orders, the Chlamydiales comprising the genus *Chlamydia* only, and the Rickettsiales comprising the other genera, on the grounds that reproduction is very different, being by binary fission inside the host cells in Rickettsiales, and including different intracellular and extracellular stages in Chlamydiales.

Summary of the classification of prokaryotes in Bergey's 8th edition (1974)

There are 19 parts which are not given taxonomic status. Some parts contain orders, others only families and genera, or genera only that cannot be more exactly assigned. Many groups include families and genera of uncertain affiliation which are not included here unless they contain one of the above examples. The examples given are those mentioned above, and the number beside each example indicates to which of the original ten orders it belonged. (Bergey uses the alternative spelling, procaryotes.)

Kingdom Prokaryotae

    Division I The Cyanobacteria (blue-green algae)

    Division II The Bacteria

        Part 1. PHOTOTROPHIC BACTERIA

        Order Rhodospirillales
        e.g. *Chlorobium* 1, *Chromatium* 1, *Rhodomicrobium* 3, *Rhodospirillum* 1.

        Part 2. THE GLIDING BACTERIA

        Order Myxobacterales
        e.g. *Archangium* 8, *Myxococcus* 8.

        Order Cytophagales
        e.g. *Beggiatoa* 7, *Cytophaga* 8, *Simonsiella* 6, *Sporocytophaga* 8, *Thiothrix* 7.

        Part 3. THE SHEATHED BACTERIA
        e.g. *Crenothrix* 2, *Leptothrix* 2, *Sphaerotilus* 2.

        Part 4. BUDDING and/or APPENDAGED BACTERIA
        e.g. *Hyphomicrobium* 3.

        Part 5. THE SPIROCHETES

        Order Spirochaetales
        e.g. *Borrelia* 9, *Leptospira* 9, *Spirochaeta* 9, *Treponema* 9.

        Part 6. SPIRAL and CURVED BACTERIA
        e.g. *Spirillum* 1.

        Part 7. GRAM-NEGATIVE AEROBIC RODS and COCCI
        e.g. *Acetobacter* 1, *Azotobacter* 4, *Pseudomonas* 1, *Rhizobium* 4.

        Part 8. GRAM-NEGATIVE FACULTATIVELY ANAEROBIC RODS
        e.g. *Escherichia* 4, *Pasteurella* 4, *Salmonella* 4, *Vibrio* 1.

        Part 9. GRAM-NEGATIVE ANAEROBIC BACTERIA
        e.g. *Bacteroides* 4, *Desulfovibrio* 1.

Part 10. GRAM-NEGATIVE COCCI and COCCOBACILLI
e.g. *Neisseria* 4.

Part 11. GRAM-NEGATIVE ANAEROBIC COCCI
e.g. *Veillonella* 4.

Part 12. GRAM-NEGATIVE, CHEMOLITHOTROPHIC BACTERIA
e.g. *Nitrobacter* 1, *Nitrosomonas* 1, *Thiobacillus* 1.

Part 13. METHANE-PRODUCING BACTERIA
e.g. *Methanobacterium* 1, *Methanococcus* 4.

Part 14. GRAM-POSITIVE COCCI
e.g. *Staphylococcus* 4, *Streptococcus* 4.

Part 15. ENDOSPORE-FORMING RODS and COCCI
e.g. *Bacillus* 4, *Clostridium* 4.

Part 16. GRAM-POSITIVE ASPOROGENOUS ROD-SHAPED BACTERIA
e.g. *Caryophanon* 6, *Lactobacillus* 4.

Part 17. ACTINOMYCETES and RELATED ORGANISMS
e.g. *Corynebacterium* 4, *Eubacterium* 4.

> Order Actinomycetales
> e.g. *Actinomyces* 5, *Mycobacterium* 5, *Streptomyces* 5.

Part 18. THE RICKETTSIAS

> Order Rickettsiales
> e.g. *Coxiella, Ehrlichia, Rickettsia.*

> Order Chlamydiales
> *Chlamydia*

Part 19. THE MYCOPLASMAS

Class Mollicutes

> Order Mycoplasmatales
> e.g. *Acholeplasma* 10, *Mycoplasma* 10.
> Genera of uncertain affiliation: *Thermoplasma* 10, *Spiroplasma* 10.
> Mycoplasma-like bodies in plants.

Bergey's suggestion for a future classification of prokaryotes

Bergey's 8th edition suggests that it may be unreasonable to separate blue-green algae into their own division, and that the differences between them and the phototrophic bacteria are at present given too much weight. It proposes the following scheme for the future:

Kingdom Prokaryotae

> Division I: Phototrophic prokaryotes (Photobacteria)

> > Class I: Blue-green photobacteria (present blue-green algae)

> > Class II: Red photobacteria ⎤ Part 1 above,
> > Class III: Green photobacteria ⎦ i.e. order Rhodospirillales

Division II: Prokaryotes indifferent to light (Scotobacteria)

Class I: The Bacteria (Parts 2–17 above)

Class II: Obligate intracellular Scotobacteria in Eukaryotic cells – Rickettsias (Part 18 above)

Class III: Scotobacteria without cell walls – Mollicutes (Part 19 above)

The differences between the three classes of phototrophic bacteria would be:

Class I Blue-green photobacteria	Class II Red photobacteria	Class III Green photobacteria
Chlorophyll *a*	Bacteriochlorophyll *a* or *b*	Bacteriochlorophyll *c* or *d*
Photosynthesis occurs under aerobic conditions and oxygen is produced	Photosynthesis occurs under anaerobic conditions and oxygen is not produced	
Photopigments on thylakoids	Photopigments on internal membrane of different structure but continuous with cytoplasmic membrane	Photopigments in Chlorobium-vesicles underlying and attached to cytoplasmic membrane
Blue-green algae	Sulphur and non-sulphur purple bacteria e.g. *Rhodospirillum, Chromatium*	Green sulphur bacteria e.g. *Chlorobium*
	order Rhodospirillales above	

The differences between the other groups have been discussed above.

4 Algae

The Algae itself may be ranked as a division, in which case it is known as Phycophyta, or the classes of algae can be considered to be so different from one another that they are raised to division (phylum) level, as is done here, and Algae becomes a group above this level (superphylum).

Algae are difficult to characterize, because they form a large amorphous group; no one characteristic is diagnostic, and many characteristics are concerned with what algae lack rather than what they possess.

The important characteristics of the Algae are:
Photosynthetic plants with a range of carotinoid and biloprotein pigments in addition to chlorophyll. Plant body unicellular, colonial, filamentous, parenchymatous or thalloid, but never forms root, stem and leaves. Lack vascular tissue. Lack archegonia. Organs producing spores or gametes lack a covering of sterile cells (except in Charophyta). Zygote never develops into multicellular embryo inside female sex organ.

The divisions of the Algae considered here are:

Cyanophyta	Pyrrophyta
Chlorophyta	Cryptophyta
Xanthophyta	Euglenophyta
Chrysophyta	Phaeophyta
Bacillariophyta	Rhodophyta

The position of the Cyanophyta (blue-green algae)

The Cyanophyta were formerly included in the Algae, and any treatise on algae still includes them; but since they are prokaryotic, they are now better included in the Schizophyta. They are described there as Schizophyceae, so that they can be compared with bacteria, but they are characterized here too, with some differences of emphasis, so that they may be compared with the other divisions of algae. The orders are described here, rather than in the previous chapter, because the classification of prokaryotes is very fluid at present, as more information about bacteria and their relatives comes to light, while the algae are more fixed, so the arrangement of groups in this chapter is more stable than that in the previous one.

Division Cyanophyta

(Myxophyta, blue-green algae)

Photosynthetic pigment is chlorophyll *a* only. Characteristic biloproteins C-phycocyanin and C-phycoerythrin present, with predominance of phycocyanin giving blue colour, but some red due to phycoerythrin. Carotinoids include *β*-carotene and unique carotinoids myxoxanthin and myxoxanthophyll. Food reserves include a special starch, cyanophycean or myxophycean starch, and a protein, cyanophycin. Cell wall composed of a peptidoglycan as in bacteria, and not found in other organisms. Flagella completely absent; if there is movement it is by gliding. Non-sexual reproduction by various types of spores, simple division, fragmentation, or in some orders by hormogonia. Sexual reproduction was thought to be unknown, but has been little investigated and there are some reports of its occurrence. Cells always small and prokaryotic, and body types include unicellular, colonial, filamentous and trichome-forming forms; cells often contain gas vacuoles. Habitats include water, especially fresh water, soil, partners in lichens, salt marshes.

Summary classification of division Cyanophyta

Class Cyanophyceae

    Order Chroococcales

    Order Chamaesiphonales (Dermocarpales)

    Order Pleurocapsales

    Order Nostocales ⎤

    Order Stigonematales ⎦ Hormogonales (Oscillatoriales)

Class Cyanophyceae

(Myxophyceae, Schizophyceae)

Characteristics as of division.

Order Chroococcales

Most unicellular or colonial in gelatinous matrix; never show trichome organization but pseudofilamentous colonies found. Reproduction by fission or colony fragmentation; endospores and hormogonia never formed. Heterocysts never found, but nannocytes sometimes present.

e.g. *Anacystis* (*Microcystis*), *Chlorogloea, Chroococcus, Entophysalis, Gloeocapsa, Synechococcus.*

Order Chamaesiphonales

(Dermocarpales)

Unicellular and colonial, and *Stichosiphon* has superficial resemblance to filament. Reproduction by endospores or exospores; hormogonia never formed. Heterocysts never found.

e.g. *Chaemaesiphon, Dermocarpa, Stichosiphon.*

Order Pleurocapsales

Body filamentous, most species heterotrichous but with reduction of prostrate or erect system. Reproduction by endospores; hormogonia never formed. Heterocysts never found.

e.g. *Pleurocapsa, Xenococcus.*

This order may be included in the Chaemaesiphonales.

Order Nostocales

Body filamentous with or without sheath, and either unbranched or with false branching. Reproduction by hormogonia. In some genera, some cells differentiated into heterocysts and akinetes.

e.g. *Lyngbya, Nostoc, Oscillatoria, Plectonema, Scytonema.*

Order Stigonematales

Body filamentous with true branching. Reproduction by hormogonia. Heterocysts in some genera.

e.g. *Haplosiphon, Mastigocladus, Stigonema.*

The Nostocales and Stigonematales are the two largest orders of blue-green algae. They are sometimes considered to be suborders of the order **Hormogonales** (Oscillatoriales), because both produce hormogonia. In some classifications the Hormogonales is raised to a higher status as **Hormogoneae**, and five filamentous orders are recognized within it as follows:

Order Oscillatoriales

Filaments unbranched. No differentiation of cells.

e.g. *Lyngbya, Oscillatoria, Trichodesmium.*

Order Nostocales

Filaments unbranched. Cells differentiated to form heterocysts and akinetes.

e.g. *Anabaena, Nostoc.*

Order Scytonematales

Filaments with false branching. Cells differentiated.
e.g. *Scytonema*, *Tolypothrix*.

Order Stigonematales

Filaments with true branching. Cells differentiated.
e.g. *Haplosiphon*, *Stigonema*.

Order Rivulariales

Filaments sometimes with false branching, but filaments taper from base to apex. Cells differentiated, with heterocysts and akinetes at base of filament.
e.g. *Calothrix*, *Rivularia*.

Division Chlorophyta

(green algae, formerly Isokontae)

Photosynthetic pigments are chlorophyll *a* and *b* as in higher plants. No biloproteins. Carotinoids include β-carotene, which is replaced by α-carotene in the Caulerpales and added to by γ-carotene and lycopene in Charophyceae; main xanthophyll is zeaxanthin. Food reserves are true starch, as in higher plants, with less fat than in many groups of algae. Cell wall usually contains cellulose, but this is replaced by xylan in Siphonales; some have no cell walls and are also classified as protozoans. 2 to 4 flagella, anterior, equal and all whiplash. Non-sexual reproduction usually by zoospores; vegetative reproduction by fragmentation. Sexual reproduction by isogamy, anisogamy or oogamy. Division large and plant body variable, may be unicellular, colonial, filamentous, or parenchymatous; uninucleate, multinucleate or coenocytic. Most in fresh water, some in soil, a few marine.

There are considered to be two or three classes of Chlorophyta, the Chlorophyceae, Charophyceae and, in some classifications, the Prasinophyceae. The latter class has an uncertain position and is not included here but is described at the end of the chapter; the Charophyceae may be raised to division level as Charophyta. In some systems other classes are introduced, namely the Oedogoniophyceae, Conjugatophyceae and Bryopsidophyceae; these classes are in square brackets in the summary below and are briefly mentioned in the text.

Summary classification of the division Chlorophyta

Class Chlorophyceae

    Order Volvocales

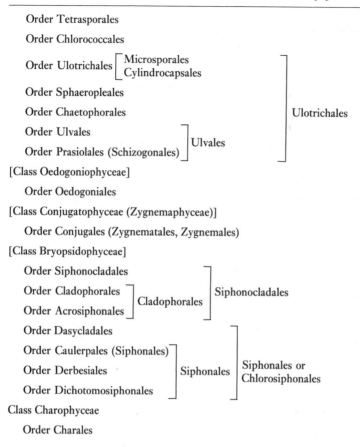

Order Tetrasporales

Order Chlorococcales

Order Ulotrichales [Microsporales / Cylindrocapsales]

Order Sphaeropleales

Order Chaetophorales

Order Ulvales

Order Prasiolales (Schizogonales)] Ulvales

Ulotrichales

[Class Oedogoniophyceae]

Order Oedogoniales

[Class Conjugatophyceae (Zygnemaphyceae)]

Order Conjugales (Zygnematales, Zygnemales)

[Class Bryopsidophyceae]

Order Siphonocladales

Order Cladophorales] Cladophorales

Order Acrosiphonales]

Siphonocladales

Order Dasycladales

Order Caulerpales (Siphonales)]

Order Derbesiales] Siphonales

Order Dichotomosiphonales]

Siphonales or Chlorosiphonales

Class Charophyceae

Order Charales

Class Chlorophyceae

Body may be unicellular, colonial, filamentous, parenchymatous or pseudo-parenchymatous, but never shows whorls of branches as in Charophyceae. Never encrusted with lime. Carotenes are β-carotene, which is replaced by α-carotene in Caulerpales. Reproductive organs not surrounded by sterile tissue. Where male gamete motile, it is not an elongate biflagellate structure. Variety of sexual and non-sexual reproduction. Zygote never germinates into protonemal filament.

Order Volvocales

Body unicellular or colonial; colonies regular and show differentiation into vegetative and reproductive cells. Vegetative stage motile in both unicellular and colonial forms. Non-sexual reproduction by simple fission or zoospores.

Sexual reproduction by isogamy, anisogamy or oogamy. Some can lose their chloroplasts and others lose their cell wall; these forms are also classified as protozoans.

e.g. *Carteria, Chlamydomonas, Chloromonas, Eudorina, Gonium, Hyaliella, Pandorina, Phacotus, Pleodorina, Polytomella, Volvox.*

Order Tetrasporales

Colonial and irregular gelatinous aggregates only. Normally non-motile, but sometimes acquire flagella; cells have capacity for vegetative cell division. Sexual reproduction by isogamy.

e.g. *Palmella, Sphaerocystis, Tetraspora.*

In some classifications this order does not exist, and its members are included in the Volvocales or Chlorococcales.

Order Chlorococcales

Body unicellular or colonial aggregates, free living or attached. Cells non-motile and have no capacity for normal vegetative cell division. Non-sexual reproduction by autospores, zoospores or aplanospores. Sexual reproduction by isogamy, anisogamy or oogamy. Many genera form secondary carotinoids, some live as part of lichens, others in soil.

e.g. *Chlorella, Chlorococcum, Scenedesmus, Trebouxia.*

Order Ulotrichales

This order is sometimes split up, with some members being placed in the Microsporales, Cylindrocapsales, Sphaeropleales, Chaetophorales, Ulvales and Prasiolales. The characteristics of the order (in the wide sense) are: Body basically filamentous with uninucleate cells. Filaments unbranched or form parenchymatous types; parenchymatous forms may be separated into the Ulvales and Prasiolales. Non-sexual reproduction by zoospores (and meiospores if *Prasiola* included here rather than in Prasiolales). Sexual reproduction by isogamy, anisogamy or oogamy. Alternation of generations isomorphic or heteromorphic.

e.g. *Cylindrocapsa, Enteromorpha, Hormidium, Microspora, Monostroma, Sphaeroplea, Ulothrix, Ulva.*

Order Microsporales

This order may be included in the Ulotrichales. When separated from them it is because:
Cell wall made of 2 segments; chloroplasts form parietal network. Sexual reproduction by isogamy or anisogamy.

e.g. *Microspora.*

Order Cylindrocapsales

This order may be included in the Ulotrichales. When separated from them it is because:
Filaments made of palmelloid cells enclosed in mucilage, with massive chloroplast. Sexual reproduction by oogamy.
e.g. *Cylindrocapsa*.

Order Sphaeropleales

This order may be included in the Ulotrichales. When separated from them it is because:
Members of Sphaeropleales have multinucleate cells (and for this reason may be placed in the class Bryopsidophyceae next to Cladophorales), and those of Ulotrichales have uninucleate cells.
e.g. *Sphaeroplea*.

Order Chaetophorales

This order may be included in the Ulotrichales. Its characteristics are:
Body consists of uniseriate branched filaments, except in *Pleurococcus* where body reduced to single cell; cells uninucleate. Fundamentally heterotrichous, but sometomes only erect part present, or body reduced to prostrate disc; branches taper and terminate in hair-like extensions. Non-sexual reproduction by macrozoospores or microzoospores, aplanospores or akinetes. Sexual reproduction by isogamy, anisogamy or oogamy.
e.g. *Aphanochaete, Cephaleuros, Chaetophora, Coleochaete, Draparnaldia, Pleurococcus* (also called *Protococcus*), *Stigeoclonium, Ulvella*.
   Some members of this order may be removed to the orders Coleochaetales and Trentepohliales.

Order Coleochaetales

Separated from Chaetophorales because sexual reproduction usually by oogamy or anisogamy.
e.g. *Aphanochaete, Chaetonema, Coleochaete*.

Order Trentepohliales

Separated from Chaetophorales because there is no tapering of branches and no hairs; sexual reproduction by isogamy.
e.g. *Cephaleuros, Trentepohlia*.

Order Ulvales

This order may be included in the Ulotrichales and is similar to the

Prasiolales:
Body parenchymatous or filamentous, and always filamentous in juvenile form, resembling *Ulothrix*. Thallus has structure like a filament that has divided in more than 1 plane; cells not arranged in groups of 4, and chloroplast variable but never large and stellate. Non-sexual reproduction by fragmentation and by quadriflagellate zoospores. Sexual reproduction by isogamy with biflagellate gametes. Isomorphic and heteromorphic alternation of generations.
e.g. *Enteromorpha* (sea guts), *Monostroma*, *Schizomeris*, *Ulva* (sea lettuce).

Order Prasiolales

(Schizogonales)
This order may be included in the Ulotrichales and is similar to the Ulvales: Body parenchymatous or only slightly more than filamentous; cells uninucleate. Distinguished from Ulvales in that thallus is a flat sheet of cells; cells small, polygonal and often arranged in groups of 4; chloroplast large and stellate with central pyrenoid. Non-sexual reproduction by akinetes. *Prasiola* has unique life history.
e.g. *Prasiola*, *Schizogonium*.

The following order has been placed in a separate class, **Oedogoniophyceae**.

Order Oedogoniales

Body consists of simple or branched filaments with a modified attachment cell; cells uninucleate. Vegetative cell division highly specialized. Non-sexual reproduction by flagellated zoospores. Sexual reproduction is an advanced oogamy with flagellated male gametes; some species of *Oedogonium* have dwarf male plants produced from special zoospores called androspores.
3 genera: *Bulbochaete*, *Oedocladium*, *Oedogonium*.

The following order has been placed in a separate class, **Conjugatophyceae** (**Zygnemaphyceae**), due to its method of reproduction, complexity of chloroplast and absence of flagella. The families within it are then raised to order status as Desmidiales, Mesotaeniales and Zygnemales. Sometimes a fourth order, **Gonatozygales**, is recognized, or may be included in the Mesotaeniales.

Order Conjugales

(Zygnematales, Zygnemales)
Body unicellular (desmids) or filamentous; cells uninucleate. There are 3 families: Desmidiaceae, true or placoderm desmids (cell wall constricted into

2 semicells), Mesotaeniaceae, saccoderm desmids (cells wall not constricted into semicells) and Zygnemataceae or Zygnemaceae (filamentous). Flagellated stages completely absent; non-sexual reproduction by fragmentation or occasionally akinetes and aplanospores in filamentous forms, and is uncommon in desmids. Sexual reproduction by conjugation, the fusion of amoeboid gametes. Chloroplast large and varied.

e.g. of placoderm desmids: *Cosmarium, Desmidium, Staurastrum.*

e.g. of saccoderm desmids: *Mesotaenium, Nectrium, Spirotaenia.*

e.g. of filamentous forms: *Mougeotia, Sirogonium, Spirogyra, Zygnema.*

The following orders have been placed in a separate class, **Bryopsidophyceae** on the grounds that they are all either coenocytic or semicoenocytic, i.e. have multinucleate cells.

Order Siphonocladales

Body consists of simple or branched filaments or vesicle-like forms, all with multinucleate cells. Vesicle forms have specialized cell division. Non-sexual reproduction by zoospores. Sexual reproduction by isogamy or anisogamy. Chloroplast net-like. May have isomorphic or heteromorphic alternation of generations, or haploid generation may be suppressed.

e.g. *Siphonocladus, Valonia.*

Order Cladophorales

This order is sometimes included in the Siphonocladales. If separated from them it is because there is no initial vesicle-like form.

They are separated from the Acrosiphonales because here the cell wall contains cellulose I and haploid and diploid stages are similar (isomorphic alternation of generations).

e.g. *Chaetomorpha, Cladophora, Pithophora.*

Order Acrosiphonales

This order is sometimes included in the Cladophorales. If separated from them it is because the cell wall here contains cellulose II, and haploid and diploid stages are dissimilar (heteromorphic alternation of generations).

3 genera: *Acrosiphonia, Spongomorpha, Urospora.*

With the Cladophorales, this order may be included in the Siphonocladales.

Order Dasycladales

Body coenocytic; uninucleate in vegetative stage, but reproductive stage consists of erect axis with whorled appendages near apex containing gametangia with many nuclei. Non-sexual reproduction absent. Sexual reproduction of diploid generation by isogamy and results in a cyst. Plastids disc-like; plant body sometimes encrusted with lime.

e.g. *Acetabularia* (mermaid's wine glass), *Dasycladus, Neomeris*.

This order is sometimes included in the Siphonales, but less commonly than the Derbesiales and Dichotomosiphonales, and has formerly been included in the Siphonocladales.

Order Caulerpales

(formerly called Siphonales, or Chlorosiphonales when including Dasycladales)

Body coenocytic, arising from rhizoid. Cell wall of mannan and xylan instead of cellulose. Non-sexual reproduction by propagules or fragmentation. Vegetative plant diploid, and sexual reproduction usually by anisogamy. In the cells, α-carotene replaces β-carotene, and there are many discoid chloroplasts whose characteristic pigment is siphonoxanthin.

e.g. *Bryopsis, Caulerpa, Codium, Halimeda*.

Order Derbesiales

This order may be considered to be part of the Caulerpales (Siphonales) or may be separated from it because its non-sexual reproduction is by zoospores and the plants show alternation of generations, the gametophytes being oval vesicles and named *Halicystis*, and the sporophytes branched and coenocytic and named *Derbesia*.

e.g. *Derbesia, Halicystis*, which have now been shown to be alternate generations of the same plant.

Order Dichotomosiphonales

This order may be considered as part of the Caulerpales (Siphonales) or separated from it because here the cells lack siphonoxanthin, and sexual reproduction is a distinct oogamy.

e.g. *Dichotomosiphon*.

Class Charophyceae

(stoneworts)

Body filamentous and consists of erect axis differentiated into nodes and internodes, with lateral branches in whorls at nodes; cells large. In some genera, especially *Chara*, cells impregnated with lime, giving common name stoneworts. Carotenes include γ-carotene and lycopene. Reproductive organs surrounded by sterile tissue. Motile male gamete present, and is an elongated biflagellated structure. Sexual reproduction a complex oogamy. Zygote germinates into protonemal structure from which adult develops.

Order Charales

Characteristics as of class.

e.g. *Chara* (stonewort), *Nitella, Protochara, Tolypella*.

This class is so different from the rest of the Chlorophyta that it is sometimes placed in a separate division, Charophyta, or may even be considered distinct from the algae altogether.

Division Xanthophyta

(yellow-green algae, formerly Heterokontae)

Photosynthetic pigments include chlorophyll *e* in a few genera as well as chlorophyll *a* (chlorophyll *e* thought unique to this division). No biloproteins. Excess of carotinoids over chlorophylls giving yellow-green colour; carotinoids include β-carotene, and xanthophylls heteroxanthin, vaucheriaxanthin and diadinoxanthin. Food reserves include oil, fat and chrysolaminarin, never starch. Cell wall frequently absent; when present contains more pectic materials than in Chlorophyta; silicification of wall common, and it is often in 2 halves in non-filamentous forms. 2 flagella present and of unequal length, the long being tinsel and the short whiplash. Non-sexual reproduction by zoospores, aplanospores, or statospores. Sexual reproduction rarely observed and is by isogamy or oogamy. Body types include unicellular, colonial, filamentous and coenocytic. Most members occur in freshwater habitats.

This division was formerly included in the Chlorophyta as the subgroup Heterokontae, having flagella of unequal length, while the other members with flagella of equal length were known as Isokontae, so Heterokontae is now a synonym for Xanthophyta and Isokontae for Chlorophyta. The division shows many parallels with the Chlorophyta.

Summary classification of the division Xanthophyta

Class Xanthophyceae

Order Heterochloridales (Choramoebales)

Order Heterogloeales (Heterocapsales)

Order Heterococcales (Mischococcales)

Order Heterotrichales (Tribonematales)

Order Heterosiphonales (Vaucheriales, Botrydiales)

Order Rhizochloridales

Class Xanthophyceae

Characteristics as of division.

Order Heterochloridales

(Chloramoebales)

All unicellular and motile by flagella; naked, with plasma membrane which is not rigid, so tend to become amoeboid. Reproduction non-sexual only, by cell division or statospores. Some species can lose chloroplasts and are also classed as protozoans.

e.g. *Chloramoeba, Heterochloris.*

Order Heterogloeales

(Heterocapsales)

Form palmelloid colonies, but individuals can revert to motile stage. Reproduction non-sexual only and is by cell division and biflagellate zoospores, and thick-walled statospores are known. Typically do not lose chloroplasts.

e.g. *Gloeochloris, Heterogloea.*

Order Heterococcales

(Mischococcales)

Cells coccoid in form, usually unicellular, and tend to attach to substratum. Membrane may be impregnated with silica. Reproduction non-sexual only and by zoospores and aplanospores. Typically do not lose chloroplasts.

e.g. *Botrydiopsis, Characiopsis, Chlorobotrys, Monodus.*

Order Heterotrichales

(Tribonematales)

Mostly filamentous; usually unbranched. Non-sexual reproduction by biflagellate zoospores. Sexual reproduction by isogamy. Typically do not lose chloroplasts.

e.g. *Monocilia, Tribonema.*

Order Heterosiphonales

(Vaucheriales, Botrydiales)

Coenocytic; the only 2 common genera are *Botrydium* which is anchored to mud by rhizoids, and *Vaucheria* which is a floating tubular thallus. Non-sexual reproduction by zoospores. Sexual reproduction by isogamy (or possibly anisogamy) in *Botrydium*, and by oogamy in *Vaucheria*.

e.g. *Botrydium, Vaucheria.*

Order Rhizochloridales

Permanently amoeboid in vegetative stage, with or without an envelope (lorica), and may tend to collection in a mass. Reproduction non-sexual only and by cell division, but in *Myxochloris* by zoospores. Can lose chlorophyll and live holozoically and are also considered protozoans.
e.g. *Chlorarachnion, Myxochloris, Rhizochloris.*

Recently a separate class of Xanthophyta, the **Eustigmatophyceae**, has been proposed, comprising a group of coccoid forms in which the stalked pyrenoids project from the inner face of the chloroplast, and the eyespot is independent of the plastid. *Chlorobotrys*, at present in the Heterococcales, should be transferred here.

Division Chrysophyta

(golden algae)

Photosynthetic pigments mainly chlorophyll *a*, although chlorophyll *c* and possibly *d* have been reported, and some members are colourless. No biloproteins. Carotinoids include β-carotene and several xanthophylls, especially fucoxanthin and diatoxanthin; appear golden because carotinoids mask chlorophyll. Food reserves are chrysolaminarin (never starch), with much oil or fat. Cellulose cell wall often absent (cells naked), but membrane may be mineralized, either silicified or calcified; naked species may be covered with scales; endogenous cysts (statospores) common. Number and type of flagella variable: cells may be uniflagellate or biflagellate, equal or unequal, similar or dissimilar. Non-sexual reproduction by cell division, zoospores and statospores. Sexual reproduction (where reported) by isogamy. Most single-celled. Very common and numerous components of phytoplankton from different habitats, chiefly fresh water; 1 family, the Coccolithophoridaceae (coccolithophorids) possess plates of calcium carbonate (coccoliths) embedded in gelatinous envelope, common in sea, and their skeletons were responsible for formation of chalk. Some species can lose chloroplasts and live holozoically and are also considered to be protozoans.

This division contains two classes, the Chrysophyceae and Haptophyceae, which are separated on the structure of the flagella, type of scales and presence of a haptonema. The classes can only be separated on details which require electron microscope study. Since this has not been done in all genera, many cannot be assigned to a class. As this distinction is very important, some authorities raise the Haptophyceae to division level as Haptophyta. The arrangement of orders within the classes is also variable as more species are examined under the electron microscope, and at present the taxonomy at this level is very fluid.

Summary classification of the division Chrysophyta

Class Chrysophyceae

    Order Chromulinales ⎤
                             Chrysomonadales
    Order Ochromonadales ⎦

    Order Chrysosphaerales

    Order Chrysocapsales

    Order Rhizochrysidales

    Order Phaeothamniales (Chrysotrichales)

Class Haptophyceae

    Order Isochrysidales

    Order Prymnesiales

Class Chrysophyceae

Possess 1 tinsel flagellum with or without a 2nd whiplash flagellum. No haptonema (see Haptophyceae). Scales, if present, mineralized with silica, of variable structure, and with no basic type of construction.

Order Chrysomonadales

Unicellular or colonial. Motile. Naked and bounded by periplast, with 1 or 2 flagella. Can lose chlorophyll and live holozoically and are also classified as flagellate protozoans.

This order can be divided into two orders:

Order Chromulinales

Characteristics of Chrysomonadales. Pyrenoids present in chloroplasts. e.g. *Chromulina*.

Order Ochromonadales

Characteristics of Chrysomonadales. Pyrenoids absent from chloroplasts. e.g. *Ochromonas*.

Order Chrysosphaerales

Coccoid in form. Non-motile. Cell wall present in some species. Do not lose chlorophyll and live holozoically. e.g. *Chrysosphaera, Epichrysis*.

Order Chrysocapsales

Palmelloid in form with cells embedded in structureless spherical or ellipsoidal mass of mucilage. Non-motile. Do not possess a cell wall. Do not lose chlorophyll and live holozoically.

e.g. *Chrysocapsa, Hydrurus.*

Order Rhizochrysidales

Rhizopodal or dendroid in form; cells amoeboid; in some genera rhizopodia are put out to anchor colony, in others delicate rhizopodia are united by coarse protoplasmic strands to form short chains of cells. Some flagellated, but flagella can be lost. Do not possess cell wall. Can lose chlorophyll and live holozoically, and are also classified as protozoans.

e.g. *Chrysamoeba, Chrysidiastrum, Rhizochrysis.*

Order Phaeothamniales

(Chrysotrichales)

Filamentous, unbranched or branched.

e.g. *Nematochrysis, Phaeodermatium, Phaeothamnion.*

Class Haptophyceae

Possess 2 whiplash flagella. Some species possess a haptonema, a 3rd thread-like appendage, apparently used for temporary anchorage of cell. Scales, if present, not mineralized with silica and consist solely of organic material and are of uniform type. Also classified as protozoans.

The coccolithophorids are now thought to belong to the Haptophyceae, since some possess a haptonema and scales, and the flagella are whiplash. The coccoliths of calcium carbonate are not mineralized scales, but are external plates embedded in a gelatinous envelope. Where scales are present, they are not silicified.

Order Isochrysidales

Flagellate stage without haptonema.

e.g. *Isochrysis.*

Order Prymnesiales

Flagellate stage with haptonema.

e.g. *Prymnesium*, and the coccolithophorids, such as *Coccolithus.*

Alternative former classifications of the Chrysophyta

Some systems do not yet recognize the classes Chrysophyceae and Haptophyceae. There are several alternative classifications.

In one type, the orders of Chrysophyceae listed above are used, with what are now considered to be Haptophyceae scattered amongst them.

Another type is the system which recognizes the Algae as of division status, and the Chrysophyceae as a class with many fewer orders. The diatoms are then considered an order of Chrysophyceae, rather than a division or class. On this classification four orders are recognized:

■ ■ ■ Order Chrysomonadales

Unicellular. Motile with 2 unequal flagella, and some members having a haptonema.
e.g. *Chromulina, Prymnesium, Synura,* coccolithophorids.

■ ■ ■ Order Chrysocapsales

Unicellular or colonial. Non-motile.
e.g. *Hydrurus.*

■ ■ ■ Order Phaeothamniales

(Chrysotrichales)

Cells form simple or branched filaments. Usually can produce flagella.
e.g. *Phaeothamnion.*

■ ■ ■ Order Diatomales

(Bacillariales)

Unicellular or colonial, with cell in 2 halves. Can produce flagella. See Bacillariophyta.

In some systems of classification, the Xanthophyceae and Bacillariophyceae are included as classes in the Chrysophyta, i.e.:

Division Chrysophyta

    Class Chrysophyceae (including Haptophyceae)

    Class Xanthophyceae

    Class Bacillariophyceae

on the grounds that they all have an excess of carotinoids over chlorophyll and appear brown.

Division Bacillariophyta

(diatoms)

Photosynthetic pigments are chlorophyll *a* and *c*. No biloproteins. Carotinoids include β- and ε-carotene, and several xanthophylls, especially fucoxanthin. Food reserves are chrysolaminarin, and much fat probably aiding buoyancy in planktonic species. Cell wall consists of pectin impregnated with silica, highly ornamented by striations and dots; cell wall called a frustule. Motile stages (gametes) possess single tinsel flagellum. Non-sexual reproduction by cell division producing cells of unequal size. Sexual reproduction by isogamy (in Pennales) and by oogamy (in Centrales) and results in formation of auxospores from zygotes, preventing continuous decrease in size. Body type unicellular or colonial, with radial or bilateral symmetry. Very widespread, in marine and freshwater plankton, in soil, in bottom flora of lakes and in formation of diatomaceous earth.

Summary classification of the division Bacillariophyta

Class Bacillariophyceae
    Order Centrales
    Order Pennales

Class Bacillariophyceae

Characteristics as of division.

Order Centrales

Diatoms with radial symmetry. Markings on frustule arranged from central point. Non-motile. Chloroplasts usually numerous and usually as minute granules. Nucleus usually occurs in lining of cytoplasm. Sexual reproduction by oogamy and sperm flagellated. Usually planktonic and marine.
e.g. *Biddulphia, Chaetoceros, Corethron, Cyclotella, Rhizosolenia.*

Order Pennales

Diatoms with bilateral symmetry. Main structural element is a spine, either a raphe or pseudoraphe. Have gliding motion which depends on raphe. Chloroplasts large, plate-like and sometimes lobed. Nucleus normally suspended in vacuole. Sexual reproduction by isogamy and gametes amoeboid. Usually freshwater plankton, epiphytes, and in soil and mud.
e.g. *Asterionella, Bacillaria, Navicula, Nitzschia, Pinnularia, Tabellaria.*

    In some systems of classification the diatoms are reduced to a class of Chrysophyta (as Bacillariophyceae) or even reduced to an order of Chrysophyceae (see above). In this case, the Pennales and Centrales are considered to be suborders and are called Pennatae and Centricae. At the other extreme, the two orders may be given class status as Pennatibacillariophyceae and Centrobacillariophyceae, within the Bacillariophyta. To avoid naming group levels, the terms are often anglicized to pennate and centric diatoms.

Division Pyrrophyta

(Pyrrhophyta, Dinophyta, dinoflagellates)

Photosynthetic pigments include chlorophyll *a* and *c*, although some members are colourless. No biloproteins. Carotinoids include β-carotene, and 2 xanthophylls unique to the group, peridinin and dinoxanthin. Food reserves include mannitol and starch. Cell wall may be present and made of cellulose, or absent and replaced by periplast; distinct sculpturing on cell surface. Flagella 2 and highly characteristic, 1 whiplash and the other band-shaped; orientation and location of flagella important taxonomic features. Non-sexual reproduction by cell division or zoospores. Sexual reproduction, where observed, usually by isogamy. Most are motile single cells, some non-motile, a few colonial, palmelloid or of short filaments. Many important members of plankton. Cells show characteristic features; nucleus large and prominent with characteristic structure; most cells contain membrane-bound pusules.

Formerly the Pyrrophyta was divided into the two classes Dinophyceae and Cryptophyceae, but now the Cryptophyceae are considered to be so different that they are placed in a division of their own, the Cryptophyta. The Pyrrophyta is now made up of two classes, the Dinophyceae and Desmophyceae; some members of each class are also classified as protozoans. Formerly the Desmophyceae and Dinophyceae were considered to be subclasses, known as Desmokontae and Dinokontae, of the class Dinophyceae.

The term dinoflagellate is used as a common name for the whole group, or also, more specifically, for the motile unicellular and usually planktonic forms of Dinophyceae.

Summary classification of the division Pyrrophyta

Class Desmophyceae

    Order Prorocentrales

Class Dinophyceae

    Order Gymnodiniales

    Order Peridiniales

    Order Dinophysidales

    Order Dinocapsales

    Order Dinotrichales

Order Dinamoebidiales (Rhizodiniales)

Order Dinococcales (Phytodiniales)

Class Desmophyceae

Cell wall, when present, has longitudinal suture dividing it into 2 valves, but does not have a transverse furrow; not divided into a number of plates. 2 flagella arise from anterior of cell; whiplash flagellum directed forwards; band-shaped flagellum bends at right angles after emerging from cell and curves round base of other flagellum. Chloroplasts usually plate-like or lobed; pyrenoids usually present. Reproduction only by longitudinal division of motile cell, and similar to division in diatoms. All unicellular. Class much smaller than Dinophyceae.

Order Prorocentrales

Characteristics as of class. Some members also classified as protozoans. e.g. *Exuviaella, Prorocentrum.*

Class Dinophyceae

Cell wall, or cell surface if wall absent, has transverse and longitudinal furrow; also divided into a number of hexagonal plates; additional spines and processes present on planktonic forms. 2 flagella arise laterally; whiplash flagellum arises in vertical furrow in various positions and trails behind cell; band-shaped flagellum emerges from cell at upper end of transverse furrow and runs around cell inside furrow. Chloroplasts usually numerous and discoid; pyrenoids absent. Non-sexual reproduction by zoospores; sexual reproduction, where observed, is usually amoeboid isogamy. Unicellular motile, non-motile and multicellular types found. Large class, of which unicelluar forms (dinoflagellates) are important constitutents of phytoplankton.

Order Gymnodiniales

(unarmoured dinoflagellates)

Unicellular motile forms without cell wall. Mostly colourless and holozoic. Reproduction by division of motile cells or of encysted cells. Also classified as protozoans.

e.g. *Gymnodium, Hemidinium.*

Order Peridiniales

(armoured dinoflagellates)

Unicellular motile forms with well defined cell wall. Usually have chloroplasts, and not usually holozoic. Non-sexual reproduction by cell division of motile or encysted cells; sexual reproduction observed in 2 genera, *Glenodinium* (isogamy with flagellated gametes) and *Ceratium* (amoeboid gametes, isogamy or anisogamy). Also classified as protozoans.

e.g. *Ceratium, Glenodinium, Gonyaulax*.

Some species are luminous, some are components of 'red tides', and some secrete toxins poisonous to fish.

Order Dinophysidales

(valved dinoflagellates, Dinophysiales)

Unicellular motile forms with cell wall of 2 valves, i.e. 2 large plates, and some smaller ones; margins of transverse furrow frequently expand into wing-like structures. Many species without chloroplasts. Reproduction by cell division.

e.g. *Dinophysis*.

Order Dinocapsales

Palmelloid forms with cell wall, i.e. non-motile cells enclosed in mucilage. Possess chloroplasts, not usually holozoic. Reproduction by gymnodinid-like zoospores. Not classified as protozoans.

e.g. *Gloeodinium, Urococcus*.

Order Dinotrichales

Filamentous. Possess chloroplasts, not usually holozoic. Reproduction by zoospores, but growth by cell division of filament. Not classified as protozoans.

e.g. *Dinoclonium, Dinothrix*.

Order Dinamoebidiales

(Rhizodiniales)

Permanently amoeboid vegetative cells. Frequently holozoic. Reproduction by gymnodinid-like zoospores. Also classified as protozoans. 1 genus: *Dinamoebidium*.

Order Dinococcales

(Phytodiniales)

Non-motile coccoid forms with cellulose cell wall without plates. Not usually holozoic. Reproduction by zoospores or aplanospores, never by cell division. Not classified as protozoans.

e.g. *Cystodinium, Pyrocystis*.

In some systems of classification only two orders are recognized, the Peridiniales (motile) and Dinotrichales (non-motile). In others, the classes may be reduced to order level and known as Desmokontales and Dinophycales or Peridiniales.

Division Cryptophyta

Photosynthetic pigments are chlorophyll *a* and *c* but some are colourless. Biloproteins phycocyanin and phycoerythrin present, probably different from those in blue-green and in red algae. Carotinoids include α- and ε-carotene. Food reserve a starch-like compound. Cells naked (no cell wall) but compressed in dorsiventral plane. 2 equal flagella, both called band-shaped. Non-sexual reproduction only, by longitudinal fission. Sexual reproduction not observed. Most unicellular and motile, many with gullet lined with trichocysts, but there are non-motile unicells, and *Bjornbergiella* forms few-celled filaments. Habitat marine and freshwater plankton.

Summary classification of the division Cryptophyta

Class Cryptophyceae

    Order Cryptomonadales

    Order Cryptococcales

Class Cryptophyceae

Characteristics as of division.

Order Cryptomonadales

Motile flagellates, many colourless and also classified as protozoans. e.g. *Chilomonas, Cryptomonas, Hemiselmis.*

Order Cryptococcales

Unicellular and non-motile, or forming very short filaments. Not usually classified as protozoans.

2 genera: *Bjornbergiella, Tetragonidium.*

The position of the Cryptophyta is uncertain. Although it was once thought to be related to the Pyrrophyta, this is no longer the case, and its position next to the Pyrrophyta in this classification does not imply a relationship between them.

Division Euglenophyta

(euglenoids)

Photosynthetic pigments are chlorophyll *a* and *b* (resembling Chlorophyceae) but some members have no chlorophyll. No biloproteins. Carotinoids include β-carotene and the xanthophylls lutein, neoxanthin, astaxanthin and antheraxanthin. Food reserve paramylum (polysaccharide). No cell wall, all naked with outer covering called periplast or pellicle. 1, 2 or 3 flagella, all tinsel, but usually only 1 emerging from invagination at anterior end; cells also have reservoir, contractile vacuole, and sometimes a red eyespot; some have wriggling (euglenoid) movement. Non-sexual reproduction by cell division although cysts are known. No good evidence of sexual reproduction. Most unicellular and flagellated, but a few form palmelloid or dendroid colonies. Habitat usually fresh or brackish water or in soil, occasionally marine. The whole group are also classified as flagellate protozoans.

Summary classification of the division Euglenophyta

Class Euglenophyceae

    Order Rhabdomonadales

    Order Heteronematales

    Order Sphenomonadales

    Order Euglenamorphales

    Order Eutreptiales

    Order Euglenales

Class Euglenophyceae

Characteristics as of division.

The orders of Euglenophyceae are rather doubtful, as the characteristic of loss of chlorophyll can be achieved with many forms by varying the external conditions.

Order Rhabdomonadales

Possess 2 flagella, 1 emergent, 2nd reduced. Do not show euglenoid movement, pellicle rigid. Lack chlorophyll and are saprophytic, never holozoic. Lack an eyespot. Unicellular only.
e.g. *Rhabdomonas, Rhabdospira*.

Order Heteronematales

Possess 1 or 2 emergent flagella which move by flickering action of tip only, giving gliding movement. Usually show euglenoid movement with non-rigid pellicle. Lack chlorophyll and are holozoic with special ingestion organelle. Lack an eyespot. Unicellular only.

e.g. *Heteronema, Peranema.*

Order Sphenomonadales

Possess 1 or 2 emergent flagella. Do not show euglenoid movement, pellicle rigid. Lack chlorophyll and are saprophytic or holozoic. Lack an eyespot. Unicellular only.

e.g. *Anisonema, Petalomonas, Sphenomonas.*

Order Euglenamorphales

Possess 3 or more flagella, all emergent. Show euglenoid movement, pellicle flexible. Chlorophyll present or absent; heterotrophic and symbiotic in gut of tadpoles. Eyespot present or absent. Unicellular only.

e.g. *Euglenamorpha, Hegneria.*

Order Euptreptiales

Possess 2 emergent flagella. Show euglenoid movement, pellicle flexible. Chlorophyll present or absent, never holozoic. Eyespot present or absent. Unicellular.

e.g. *Distigma, Eutreptia.*

Order Euglenales

Possess 2 flagella, but only 1 emergent and locomotory, other non-emergent. Some show euglenoid movement, pellicle flexible. Some possess chlorophyll, others do not, but chlorophyll can easily be lost under certain environmental conditions and they will survive saprophytically. Eyespot present or absent. As well as unicellular forms, there are dendroid or palmelloid colonies.

e.g. *Astasia, Colacium, Euglena.*

The immobile dendroid and palmelloid forms, like *Colacium*, may be placed in a separate order, **Euglenocapsales (Colaciales)**.

In some classifications only two orders are recognized; photosynthetic and saprophytic forms are placed in the **Euglenales**, holozoic forms in the **Peranematales**.

Division Phaeophyta

(brown algae)

Photosynthetic pigments are chlorophyll *a* and *c*. No biloproteins. Carotinoids include β-carotene and several xanthophylls, particularly fucoxanthin and diatoxanthin; brown because fucoxanthin masks chlorophyll. Food reserves include laminarin (polysaccharide) and mannitol; many cells contain small colourless vesicles called fucosan vesicles; not known whether fucosan is a food reserve or waste product of metabolism. Cell wall contains alginic acid and fucinic acid. Motile phase consists of pear-shaped bodies with 2 unequal lateral flagella, 1 whiplash, 1 tinsel. Non-sexual reproduction by zoospores, aplanospores, monospores or tetraspores. Sexual reproduction by isogamy, anisogamy or oogamy. Plant body shows greater complexity than in any other group, except red algae, and includes minute discs, branched filaments, erect tufted uniseriate forms, thalli, multiaxial forms which are complex and often heterotrichous. Most in sea, a few in fresh water.

Summary classification of the division Phaeophyta

Class Phaeophyceae

  Isogeneratae

    Order Ectocarpales

    Order Tilopteridales

    Order Sphacelariales

    Order Dictyotales

    Order Cutleriales

  Heterogeneratae

    Order Chordariales ⎤

    Order Sporochnales ⎬ Haplostichineae

    Order Desmarestiales ⎦

    Order Dictyosiphonales ⎤

    Order Punctariales ⎬ Polystichineae

    Order Laminariales ⎦

  Cyclosporeae

    Order Fucales ⎤

    Order Durvilleales ⎬ Fucales

    Order Ascoseirales ⎦

The terms used here do not always correspond to the ICBN, but they are useful in forming natural groups of orders.

Class Phaeophyceae

Characteristics as of division.

Isogeneratae

Plants usually with isomorphic alternation of generations.

Order Ectocarpales

Body composed of uniseriate heterotrichous filaments; prostrate or erect part sometimes absent. Growth intercalary. Non-sexual reproduction by zoospores. Sexual reproduction by isogamy or sometimes anisogamy. Isomorphic alternation of generations.
e.g. *Ectocarpus, Giffordia, Phaeostroma, Pilayella, Streblonema.*

Order Tilopteridales

Body of branched uniseriate, later multiseriate heterotrichous filaments which arise from basal disc. Growth intercalary and usually trichothallic. Non-sexual reproduction by aplanospores called monospores produced singly in globose sporangia. Sexual reproduction by oogamy. Alternation of generations probably isomorphic.
e.g. *Haplospora, Tilopteris.*

Order Sphacelariales

Body composed of uniseriate, later multiseriate, heterotrichous filaments giving parenchymatous appearance. Growth by an apical cell. Non-sexual reproduction by zoospores; vegetative reproduction by propagules or fragmentation. Sexual reproduction by isogamy or physiological anisogamy. Generally isomorphic alternation of generations.
e.g. *Cladostephus, Halopteris, Sphacelaria.*

Order Dictyotales

Body a parenchymatous thallus with a pad-like holdfast, ribbon-like or fan-shaped. Growth by apical cell or marginal growing point. Non-sexual reproduction by large non-motile spores called tetraspores, not analogous to those of red algae. Sexual reproduction by oogamy. Alternation of generations usually isomorphic.
e.g. *Dictyopteris, Dictyota, Padina* (peacock's tail).

Order Cutleriales

Body a branched, flat, ribbon-like or fan-like parenchymatous thallus. Growth trichothallic. Non-sexual reproduction by zoospores. Sexual reproduction by anisogamy. Alternation of generations may be isomorphic, heteromorphic or having the non-sexual generation only, but members without isomorphic life cycle have all other characteristics of the order. e.g. *Cutleria, Microzonia, Zanardinia*.

Heterogeneratae

Plants usually with heteromorphic alternation of generations.

The Heterogeneratae may be divided into two groups: haplostichous forms (Haplostichineae or Haplophycidae) in which the plant body is pseudoparenchymatous, formed by the aggregation of filaments, and polystichous forms (Polystichineae or Polyphycidae) which are parenchymatous.

Order Chordariales

Body consists of 1 or more axial branched filaments arising from a prostrate basal portion, forming pseudoparenchymatous thallus, but can be modified or reduced. Growth commonly trichothallic, occasionally apical. Non-sexual reproduction by zoospores. Sexual reproduction by isogamy, rarely anisogamy. Alternation of generations heteromorphic and irregular, usually consisting of macroscopic sporophyte and microscopic gametophyte, some with sporophyte generation only.
e.g. *Chordaria, Leathesia* (sea potato), *Spermatochnus*.

Order Sporochnales

Body a pseudoparenchymatous thallus. Growth intercallary and trichothallic, with growing point at base of terminal tuft of hairs. Non-sexual reproduction with zoospores. Sexual reproduction by oogamy. Alternation of generations heteromorphic with gametophyte microscopic.
e.g. *Carpomitra, Sporochnus*.

Order Desmarestiales

Body a bushy pseudoparenchymatous thallus, having axis with pinnately arranged branches terminating in branched uniseriate filaments. Growth trichothallic. Non-sexual reproduction by zoospores. Sexual reproduction by oogamy. Heteromorphic alternation of generations with macroscopic sporophyte and microscopic gametophyte.
e.g. *Arthrocladia, Desmarestia*.

Order Dictyosiphonales

Body a parenchymatous thallus which may be solid, tubular, saccate or leaf-like. Growth intercalary or by apical cell. Non-sexual reproduction by zoospores. Sexual reproduction by isogamy. Alternation of generations heteromorphic with gametophyte forming dwarf generation on leafy thallus. e.g. *Dictyosiphon, Punctaria, Soranthera.*

Some members of the Dictyosiphonales may be separated into the order Punctariales. The differences between the two orders are:

Punctariales	Dictyosiphonales
Growth intercalary	Growth by single spical cell
Sporangia uni- and plurilocular	Sporangia usually unilocular
Sori over whole thallus	Sori just beneath surface
e.g. *Punctaria, Soranthera*	e.g. *Dictyosiphon*

Order Laminariales

(kelps)

Body usually a parenchymatous thallus, having distinct holdfast, stipe and lamina. Growth intercalary. Non-sexual reproduction by zoospores in sporangia in sori on frond or on special leaflets. Sexual reproduction by oogamy. Alternation of generations heteromorphic, with macroscopic, sometimes very large sporophyte, and microscopic dioecious gametophyte. e.g. *Alaria, Chorda* (sea lace), *Laminaria* (oarweed), *Macrocystis, Nereocystis, Pelagophycus, Saccorhiza* (furbellows).

Cyclosporeae

Plants with no alternation of generations but with diploid vegetative stage and haploid gametes.

Order Fucales

(Cyclosporales)

Body consists of thallus usually differentiated into holdfast, stipe and fronds. Growth usually by single apical cell. Non-sexual reproduction absent (except occasional fragmentation). Sexual reproduction by oogamy with sex organs in localized sunken regions; oogonia borne directly on wall of receptacles. Embryo has apical hairs and groove. e.g. *Ascophyllum* (wracks), *Bifurcaria, Cystoseira, Fucus* (wracks), *Himanthalia* (thong weed), *Pelvetia* (wracks), *Sargassum* (japweed).

The following two orders may be included in the Fucales.

Order Durvilleales

Body consists of thallus with large basal pad, short stipe and long dissected fronds. Growth region diffuse or marginal. Non-sexual reproduction absent. Sexual reproduction by oogamy, with sex organs in sunken regions scattered all over thallus; oogonia on branched filaments. Embryo lacks apical hairs and groove.

2 genera: *Durvillea, Sarcophycus.*

Order Ascoseirales

Body consists of a split lamina, short stipe and pad-like holdfast; thallus has inner layer like Laminariales, but outer layers like Fucales. Growth intercalary. Reproduction non-sexual only, by chains of sporangia borne in sunken regions.

1 genus and species: *Ascoseira mirabilis*, of Antarctic.

Division Rhodophyta

(red algae)

Photosynthetic pigments are chlorophyll *a* and *d*. Unique biloproteins, R-phycocyanin and R-phycoerythrin present, of different structure to those of Cyanophyta and Cryptophyta. Carotinoids include α- and β-carotene and several xanthophylls especially taraxanthin. Food reserves include floridean starch and the galactoside floridoside, and sometimes mannitol or mannoglycerate; fats present are more saturated fats. Cell wall components include polysulphate esters as well as cellulose and pectic materials. No flagellated stages (only group of eukaryotic algae with none). Sexual reproduction a highly specialized characteristic oogamy. Non-sexual reproduction by monospores or occasionally endospores and akinetes, and by cell division in unicellular forms. Thallus usually a branched filament or multifilamentous axis, but there are unicellular, simple filamentous and parenchymatous types. Mainly marine.

Summary classification of the division Rhodophyta

Class Rhodophyceae

    Subclass Bangiophycidae (Bangioideae, Protoflorideae)

        Order Porphyridiales

        Order Goniotrichales

        Order Compsopogonales   Goniotrichales   Bangiales

        Order Rhodochaetales

        Order Bangiales

    Subclass Florideophycidae (Florideae, Euf123florideae)

        Order Nemalionales

        Order Bonnemaisoniales   Nemalionales

        Order Gelidiales

Order Cryptonemiales

Order Gigartinales

Order Rhodymeniales

Order Ceramiales

Class Rhodophyceae

Characteristics as of division.

Subclass Bangiophycidae

(Bangioideae, Protoflorideae)

Body unicellular, simple or branched filaments, solid cylinders or flattened sheets of cells forming true parenchyma, but never show aggregation of filaments into pseudoparenchymatous types. Growth diffuse. Cells normally have single axile stellate chloroplast, but shape of chloroplast more variable than in Florideophycidae. Pit and protoplasmic connections between cells less common than in Florideophycidae. Life cycle comparatively simple (see diagram). Production of carpospores is by direct division of zygote. Sexual reproductive apparatus unspecialized. Gametophyte dominant. Non-sexual reproduction by monospores or occasionally by endospores or akinetes.

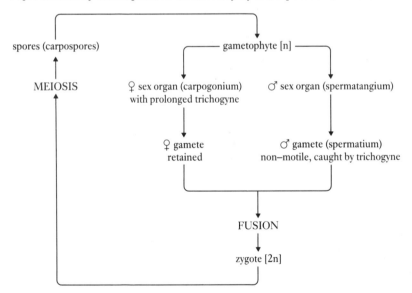

The orders of Bangiophycidae are variable depending on authority. Some authorities recognize only one order, Bangiales, with three families, while others recognize three orders according to body form:

Porphyridiales (unicellular)
Goniotrichales (filamentous)
Bangiales (parenchymatous)

Others recognize five orders, with three filamentous orders:

Porphyridiales (unicellular)
Goniotrichales ⎤
Compsopogonales ⎢ (filamentous)
Rhodochaetales ⎦
Bangiales (parenchymatous)

This is the classification used here.

Order Porphyridiales

Body unicellular, but often aggregated into gelatinous colonies. Reproduction by cell division in all planes, by monospores or endospores.
e.g. *Porphyridium*.

Order Goniotrichales

Body filamentous, with uniseriate, multiseriate, simple or branched filaments. Reproduction by monospores or akinetes.
e.g. *Asterocystis, Goniotrichum*.

Order Compsopogonales

Body filamentous, with branched filaments with axis of large cells surrounded by cortex of small, highly pigmented cells. Reproduction by monospores.
1 genus: *Compsopogon*.

Order Rhodochaetales

Body filamentous with minute branched filaments, with apical growth, and pits between cells. Non-sexual reproduction by monospores; sexual reproduction simple.
1 genus: *Rhodochaete*.

Order Bangiales

Body parenchymatous with thin fronds, or with uniseriate, later multiseriate filaments. Non-sexual reproduction by monospores; sexual reproduction simple; apparent alternation of heteromorphic haploid generations in some species.
e.g. *Bangia, Porphyra* (laver).

Subclass Florideophycidae

(Florideae, Euflorideae)

Body consists of branched filaments or pseudoparenchymatous or membranous thallus; many heterotrichous. Growth by apical cell. Cells normally have a number of parietal chloroplasts, shape of chloroplast more constant than in Bangiophycidae. Pit and protoplasmic connections more common. Life cycle more complex than in Bangiophycidae, and usually includes 3 generations, a haploid gametophyte, a diploid carposporophyte and a diploid tetrasporophyte, a kind of life cycle unknown in any other group of plants (see diagram). Production of carpospores indirect from zygote. Sexual reproduction specialized. Alternation of generations isomorphic, or heteromorphic in simpler forms. Non-sexual reproduction by monospores.

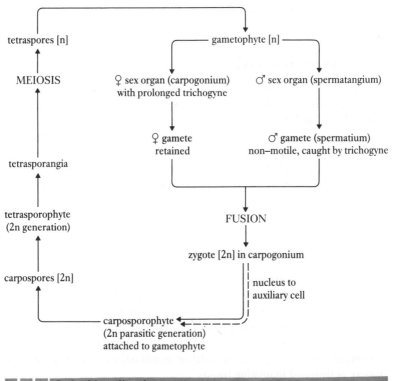

Order Nemalionales

Not 2 diploid stages, but usually zygote undergoes meiosis and germinates into haploid carposporophyte. Auxiliary cells probably absent. Alternation of generations heteromorphic or isomorphic. Body form simple, uniaxial or multiaxial, often not compact.

e.g. *Batrachospermum* (frog spawn alga), *Lemanea, Nemalion, Scinaia*.

Order Bonnemaisoniales

Division of zygote not meiotic. Auxiliary cell develops from cell of carpogonial branch and is mainly nutritive. Alternation of generations isomorphic or heteromorphic; sometimes only sexual generation present. Thallus compact, and uniaxial or multiaxial.

e.g. *Asparagopsis, Bonnemaisonia.*

This order may be included in the Nemalionales.

Order Gelidiales

2 diploid stages, carposporophyte and tetrasporophyte, alternating with haploid gametophyte. Auxiliary cell considered absent or nutritive only. Carpogonial filaments 1-celled, and cell develops into carpogonium. Alternation of generations isomorphic. All uniaxial and compact.

e.g. *Gelidium, Pterocladia.*

Order Cryptonemiales

2 diploid stages, carposporophyte and tetrasporophyte, alternating with haploid gametophyte. Auxiliary cells present, and differentiate before sexual fusion; borne on special branches. Carpogonial branches on special accessory filaments or grouped into sori; after fertilization, there are special outgrowths called ooblasts, down which zygote nucleus passes. Alternation of generations isomorphic. Uniaxial or multiaxial, erect or encrusting on surfaces; 1 family, Corallinaceae (coralline algae) heavily lime-encrusting and important rock-formers.

e.g. *Corallina, Dudresnaya, Jania, Lithothamnion.*

Order Gigartinales

2 diploid stages, carposporophyte and tetrasporophyte, alternating with haploid gametophyte. Auxiliary cells present and differentiate before sexual fusion, but are intercalary cells of vegetative thallus or supporting cells of carpogonial filament. Carpogonial filaments arise from normal vegetative thallus, not specialized accessory filaments. Alternation of generations isomorphic. Thalli based on uniaxial or multiaxial construction and often appear as flattened branching fronds.

e.g. *Ahnfeltia* (lady's wig), *Chondrus* (carragheen), *Gigartina.*

Order Rhodymeniales

2 diploid stages, carposporophyte and tetrasporophyte, alternating with haploid gametophyte. Auxiliary cells present and differentiate after sexual

fusion. But a small immature auxiliary cell, indistinguishable from vegetative cell is formed *before* fusion. Alternation of generations isomorphic. All species multiaxial; in many genera mature thallus has hollow centre divided by septa at intervals.

e.g. *Champia, Chylocladia, Gastroclonium, Lomentaria, Rhodymenia* (dulse).

Order Ceramiales

2 diploid stages, carposporophyte and tetrasporophyte, alternating with haploid gametophyte. Auxiliary cells present and differentiate after sexual fusion. But auxiliary cell borne directly on supporting cell of carpogonial filament and is not cut off until *after* fusion; carpogonial filament always 4-celled. Alternations of generations isomorphic. Vegetative form variable, includes branched filaments and pseudoparenchymatous thalli forming leafy expanses, and cartilaginous and polysiphoneous forms.

e.g. *Ceramium, Delesseria, Laurencia, Polysiphonia.*

Algae of uncertain taxonomic position

The divisions Prasinophyta and Chloromonadophyta do not fit into the above system, although, as a class, Prasinophyceae has been placed in the Chlorophyta. They are considered here as true divisions.

Division Prasinophyta

Product of photosynthesis sometimes mannitol. Unicellular with 1, 2 or 4 flagella which have 1 or 2 layers of plate-like scales.

Class Prasinophyceae

Characteristics as of division.

Order Pyramimonadales

Reproduction usually by fission; vegetative stage motile.

e.g. *Micromonas, Nephroselmis, Platymonas, Pyramimonas.*

Micromonas may be placed in a separate class, **Loxophyceae**, because scales are arranged differently.

Order Halosphaerales

Reproduction by aplanospores, cysts or zoospores; vegetative stage non-motile.

e.g. *Halosphaera.*

Division Chloromonadophyta

Pigments formerly thought to be as Xanthophyta but now uncertain; some colourless. Food reserve oil. Unicellular; no cell wall, soft periplast; cells can change shape and contain trichocysts. Usually biflagellate. Also classified as protozoans.

Class Chloromonadophyceae

Characteristics as of division.

Order Chloromonadales

Characteristics as of division.
e.g. *Gonyostomum, Merotrichia, Monomastix, Thermomastix.*

5 Fungi

Although fungi are often placed in the division Mycota or Mycophyta, authorities differ in opinion on the status of the subgroups, which has led to some confusion of terminology. There are two very different groups of fungi, the slime moulds or slime fungi, which have been classed as animals known as Mycetozoa, and the true fungi. These groups may be raised to division status as the division Myxomycota (slime moulds or slime fungi) and Eumycota (true fungi), giving the Fungi the status of a kingdom. Alternatively, they may be considered as subdivisions of fungi and known as Myxomycotina and Eumycotina, or the slime moulds may be reduced to merely a class, Myxomycetes, equivalent to the other classes of fungi such as Asco- and

Table 5.1: Alternative systems of classification of Fungi

Kingdom Fungi	Division Fungi (Mycota)
Division Myxomycota (slime moulds)	Subdivision Myxomycotina (slime moulds)
Class Acrasiomycetes	
Class Hydromyxomycetes	
Class Myxomycetes	Class Myxomycetes
Class Plasmodiophoro-mycetes*	
Division Eumycota (true fungi)	Subdivision Eumycotina (true fungi)
Subdivision Mastigomycotina	Class Chytridiomycetes ⎤
Class Chytridiomycetes	Class Hyphochytridio-
Class Hyphochytridiomycetes	mycetes
Class Oomycetes	Class Oomycetes Phycomycetes
Class Plasmodiophoro-mycetes*	Class Plasmodiophoro- (algal or mycetes* lower fungi)
Subdivision Zygomycotina	
Class Zygomycetes	Class Zygomycetes
Class Trichomycetes	Class Trichomycetes ⎦
Subdivision Ascomycotina	Class Ascomycetes ⎤ higher fungi
Subdivision Basidiomycotina	Class Basidiomycetes ⎦
Subdivision Deuteromycotina (Fungi Imperfecti)	Class Deuteromycetes (Fungi Imperfecti)

*This class may be placed in either of these positions.

Basidiomycetes. Where true fungi are considered to be of division level, the classes are raised to subdivision level, e.g. as Asco- and Basidiomycotina.

The alternative systems are outlined in Table 5.1.

The system described here includes material from both classifications but is based on that in the left-hand column, because it is more comprehensive and includes more terms that the reader is likely to meet. The simpler system of the right-hand column is still widely used, and the terms ascomycetes and basidiomycetes have become common names for the groups, even if they are formally given the status of Asco- and Basidiomycotina.

Kingdom (or Division) Fungi

(Mycota, Mycophyta)

Method of nutrition heterotrophic, lack chloroplasts. Usually possess cell walls of cellulose and/or chitin, although in slime moulds walls are only found in spores. Reproduce by spores produced sexually or non-sexually. Plant body unicellular, an amoeboid mass, or filamentous. Habitat freshwater or terrestrial, rarely marine. Food reserves glycogen, fat, mannitol and other substances, never starch.

Division Myxomycota

(slime moulds, slime fungi)

Vegetative stage not walled. Vegetative stage amoebae, pseudoplasmodium or plasmodium. Nutrition saprophytic, or holozoic for which slime moulds have been thought to be animals. Spores produced in or on sporocarps. Spores walled with keratin or cellulose, never chitin (unless Plasmodiophoromycetes included here). Sexual reproduction by fusion of compatible myxamoebae or of myxoflagellatae. Non-sexual reproduction by fission of individual cells. All terrestrial.

Summary classification of the division Myxomycota

Class Acrasiomycetes
    Order Acrasiales
Class Hydromyxomycetes
    Order Hydromyxales
    Order Labyrinthulales
Class Myxomycetes
    Subclass Ceratiomyxomycetidae
    Order Ceratiomyxales

Subclass Myxogastromycetidae

 Order Protosteliales

 Order Liceales

 Order Echinosteliales

 Order Trichiales

 Order Stemonitales

 Order Physarales

Class Plasmodiophoromycetes (for details see true fungi)

Class Acrasiomycetes

Vegetative stage free-living amoebae which unite to form a pseudoplasmodium before reproduction.

Order Acrasiales

Characteristics as of class.
e.g. *Acytostelium, Dictyostelium, Guttulina, Sappinia.*

Class Hydromyxomycetes

Vegetative stage amoebae, pseudoplasmodium or net-like plasmodium.

Order Hydromyxales

1 phase to life history.
e.g. *Plakopus, Vampyrella.*

Order Labyrinthulales

2 phases to life history.
e.g. *Labyrinthula.*

Class Myxomycetes

Vegetative stage a free-living saprophytic plasmodium.

Subclass Ceratiomyxomycetidae

Spores borne externally. Hypothallus conspicuous. On germination spores each give rise to a protoplast that cleaves into 8 myxoflagellatae.

Order Ceratiomyxales

Characteristics as of class.
1 genus: *Ceratiomyxa*.

Subclass Myxogastromycetidae

Spores borne internally. Hypothallus inconspicuous. On germination, each spore gives rise to 1 to 4 myxamoebae or myxoflagellatae.

Order Protosteliales

Sporophore minute, usually consisting of single spore on non-cellular stalk.
e.g. *Protostelium, Schizoplasmodium*.

Order Liceales

Sporophores have no capillitium, may have pseudocapillitium. Sporophores stalked or sessile. Lime absent. Spores in mass colourless, pallid or dingy.
e.g. *Licea, Reticularia*.

Order Echinosteliales

Capillitium scanty (if present) and arises from top of columella. Sporophores stalked. Lime absent. Spores in mass white, pale pink, pale yellow or brown, with conspicuously thickened areas on walls.
1 genus: *Echinostelium*.

Order Trichiales

Capillitium abundant as network or elaters. Sporophores stalked or sessile with persistent peridium and no columella. Lime rarely present. Spores in mass bright coloured.
e.g. *Arcyria, Hemitrichia, Trichia*.

Order Stemonitales

Capillitium present. Sporophores usually stalked, sometimes sessile, with columella. Lime rarely present, then confined to hypothallus, stalk and columella. Spores in mass black, violaceous or rust-coloured.
e.g. *Comatricha, Lamproderma, Stemonitis*.

Order Physarales

Capillitium present. Sporophores stalked or sessile, with peridium. Lime

characteristically present on peridium and capillitium. Spores in mass black, violaceous or rust-coloured.
e.g. *Diderma, Didymium, Fuligo, Physarum.*

Class Plasmodiophoromycetes

Vegetative stage a plasmodium parasitic within cells of host plant. The spores have walls of chitin, so these may be considered a class of true fungi, and were formerly a group of Phycomycetes. Details are described under true fungi.

Division Eumycota

(true fungi)

Vegetative stage surrounded by cell wall of cellulose and/or chitin. Vegetative stage typically filamentous (a mycelium) but occasionally unicellular. Nutrition parasitic, saprophytic or symbiotic, but not holozoic; those fungi which trap animals are actually parasitic because they put out haustoria into the prey. Spores produced in or on sporocarps which are not usually made from whole vegetative plant body. Spores walled with chitin or cellulose. Sexual reproduction by various means, usually resulting in resting structures (e.g. zygospores) or meiospores (e.g. ascospores and basidiospores). Non-sexual reproduction by fission, budding, fragmentation or more usually by spores.

The lower fungi

The so-called lower fungi were formerly placed in a class called Phycomycetes, which included the first six classes described below. It is still used as a general common name for the group of lower fungi. Its characteristics are:

Phycomycetes

(lower fungi, algal fungi)

Spores produced in sporangia. Hyphae usually coenocytic. Diploid stage usually restricted to zygote.

In the modern system, the old Phycomycetes orders are placed in two subdivisions, the Mastigomycotina and Zygomycotina. The relationship between the old and new systems is given in Table 5.2.

Subdivision Mastigomycotina

Body unicellular or a mycelium. Spores and/or gametes motile.

Table 5.2: Comparison of classifications of Phycomycetes

Present classification	Former classification
	Class Phycomycetes
Subdivision Mastigomycotina	—
Class Chytridiomycetes	—
Order Chytridiales	Order Chytridiales
Order Harpochytridiales	—
Order Blastocladiales	Order Blastocladiales
Order Monoblepharidales	Order Monoblepharidales
Class Hyphochytridiomycetes	—
Order Hyphochytriales	Order Hyphochytriales
Class Plasmodiophoro-mycetes*	—
Order Plasmodiophorales	Order Plasmodiophorales
Class Oomycetes	Order Oomycetales
Order Lagenidiales	
Order Saprolegniales	Family Saprolegniaceae
Order Leptomitales	
Order Peronosporales	Family Peronosporaceae
Subdivision Zygomycotina	—
Class Zygomycetes	Order Zygomycetales
Order Mucorales	Family Mucoraceae
Order Entomophthorales	Family Entomophthoraceae
Order Zoopagales	Family Zoopagaceae
Class Trichomycetes	
Order Amoebidiales	
Order Eccrinales	
Order Asellariales	
Order Harpellales	

*This class is also considered to be a group of slime moulds.

Class Chytridiomycetes

1 whiplash flagella directed to rear (opisthocont) in motile forms. Non-sexual reproduction by flagellated zoospores. Sexual reproduction by flagellated gametes. Body unicellular or filamentous; hyphal walls of chitin and cellulose. Holocarpic or eucarpic. Mostly aquatic, saprophytic, or parasitic on algae, fungi or flowering plants.

Order Chytridiales

Very primitive, often with no true mycelium, only naked protoplast, others with simple thallus, some advanced forms with primitive mycelium. Hyphal walls, when present, of chitin. Saprophytes, or parasites generally in lower plants or animals, a few in higher plants; some attack and capture unicellular organisms by rhizoidal processes or very fine hyphae. Nuclear cap rich in RNA lies alongside nucleus but disappears during nuclear division. Non-sexual reproduction by fission or zoospores. Sexual reproduction observed in only a few forms and is by isogamy with gametes identical to zoospores, or with gametangia.

e.g. *Chytridium, Olpidium, Polyphagus, Synchytrium.*

Order Harpochytriales

(Harpochytridiales)

Freshwater saprophytes.

2 genera: *Harpochytrium, Oedogoniomyces.*

Order Blastocladiales

Some as simple as Chytridiales, but others have richly branched, multinucleate hyphae. Most have hyphal walls of chitin. Most saprophytic in water (water moulds). Nuclear cap present as in Chytridiales. Non-sexual reproduction by zoospores, and meiospores produced in thick-walled resting sporangia. Sexual reproduction by isogamy and anisogamy.

e.g. *Allomyces, Blastocladia, Coelomomyces.*

Order Monoblepharidales

Mycelium more extensive than in previous orders, and made of branched multinucleate hyphae and can appear foamy. Hyphal walls of cellulose. Saprophytic in water (water moulds). Nuclear cap absent. Non-sexual reproduction by uniflagellate zoospores produced in club-shaped sporangia. Sexual reproduction by complex oogamy.

e.g. *Monoblepharis.*

Class Hyphochytridiomycetes

1 tinsel flagellum projecting forwards (acrocont) in motile forms. Otherwise similar to Chytridiomycetes, but mostly marine parasites on algae or other fungi; some freshwater or saprophytic.

Order Hyphochytriales

(Anisochytridiales)

Characteristics as of class.

e.g. *Anisolpidium, Hyphochytrium, Rhizidiomyces.*

Class Plasmodiophoromycetes

2 anterior, unequal whiplash flagella in motile forms. Non-sexual reproduction by zoospores. Sexual reproduction by biflagellate gametes. Body a plasmodium living inside plant tissue with no cell walls, but spore walls are of chitin. Plasmodium becomes converted to zoosporangia, or many small, walled spores. Plant parasites causing abnormal growth.

Order Plasmodiophorales

Characteristics as of class.
e.g. *Plasmodiophora, Spongospora*.
    This class and order may be considered to be a group of slime moulds since a plasmodium is produced, or placed here since spore walls are of chitin.

Class Oomycetes

2 flagella, 1 whiplash and 1 tinsel in motile forms. Non-sexual reproduction by biflagellate zoospores (absent in some terrestrial species where zoosporangium functions as conidium). Sexual reproduction results in thick-walled oospore. Body unicellular, or a filamentous mycelium with coenocytic hyphae, and hyphal walls of cellulose, not chitin. Holocarpic or eucarpic. Aquatic or terrestrial; saprophytic or parasitic.

Order Lagenidiales

Water moulds. Unicellular or filamentous. Hyphae without constrictions or cellulin plugs. Saprophytic, or parasitic on algae or fungi. Holocarpic. Oogonium contains a single ovum.
e.g. *Lagenidium, Lagenocystis, Myzocytium*.

Order Saprolegniales

Water moulds or soil fungi. Hyphae without constrictions or cellulin plugs. Saprophytic or parasitic. Mostly eucarpic. Oogonium contains 1 to many free ova. Some species diplanetic.
e.g. *Aphanomyces, Saprolegnia*.

Order Leptomitales

Aquatic, found in polluted waters. Filamentous. Hyphae have constrictions frequently with cellulin plugs. Saprophytic. Eucarpic. Oogonium contains a single ovum.
e.g. *Leptomitus* (sewage fungus).

Order Peronosporales

(downy mildews)

Aquatic or terrestrial. Filamentous. Hyphae do not have constrictions or cellulin plugs. Parasitic on algae, or more usually on vascular plants, the latter obligate parasites causing downy mildews. Eucarpic. Oogonia and antheridia do not have sharply differentiated gametes. Zoosporangia in advanced species borne on well differentiated sporangiophores and behave as conidia.
e.g. *Peronospora*, *Phytophthora* (potato blight), *Plasmopara* (grape mildew), *Pythium*.

Subdivision Zygomycotina

Body a mycelium, usually coenocytic. Spores and/or gametes non-motile. Zygospores typically produced.

Class Zygomycetes

Non-sexual reproduction by aplanospores or conidia. Sexual reproduction by conjugation resulting in production of zygospores. Eucarpic. Terrestrial; parasitic on plants, animals or people, or saprophytic.

Order Mucorales

(bread moulds)

Mostly saprophytic, some weakly parasitic on plants, some parasitic on people causing a pulmonary infection, mucomycosis. Non-sexual reproduction by sporangiospores, 1-spored sporangiola, or conidia.
e.g. *Mucor* (pin mould), *Phycomyces*, *Pilobolus* (sporangium forcibly discharged).

Order Entomophthorales

Parasitic on insects or desmids, or saprophytic. Non-sexual reproduction by modified sporangia, functioning as conidia. Spores (conidia) sometimes forcibly discharged.
e.g. *Basidiobolus*, *Entomophthora*.

Order Zoopagales

Parasitic on small animals such as amoeba, rotifers and nematodes, which they trap by various specialized mechanisms. Non-sexual reproduction by conidia borne singly or in chains. Spores not forcibly discharged.
e.g. *Cochlonema*, *Stylopage*, *Zoopage*.
    This order may be included in the Entomophthorales.

Class Trichomycetes

Non-sexual reproduction by sporangiospores, trichospores, arthrospores or amoeboid cells. Sexual reproduction (where known) as in Zygomycetes. Body a filament attached to lining of digestive tract or external cuticle of arthropods. Eucarpic. Commensals or parasites on arthropods.

Order Amoebidiales

Hyphae coenocytic. Non-sexual reproduction by amoeboid cells.
e.g. *Amoebidium, Paramoebidium.*

Order Eccrinales

Hyphae coenocytic. Non-sexual reproduction by aplanosporangia.
e.g. *Astreptonema (Eccrinella), Enterobryus (Eccrina).*

Order Asellariales

Hyphae septate, branched. Non-sexual reproduction by arthrospores.
e.g. *Asellaria.*

Order Harpellales

Hyphae septate, simple or branched. Non-sexual reproduction by trichospores. Sexual reproduction known and as in Zygomycetes.
e.g. *Genistella, Harpella.*

The higher fungi

These fungi were formerly placed in two classes, the Ascomycetes and Basidiomycetes. These terms are still retained as common names, but the groups are now given subdivision status as Ascomycotina and Basidiomycotina.

The subdivision Deuteromycotina comprises those forms which have no sexual stages, so cannot be assigned to any other group, but are thought to be non-sexual or vegetative forms of asco- or basidiomycetes.

Subdivision Ascomycotina

(ascomycetes, sac fungi)

Saprophytic, or parasitic on plants, animals and people. Occasionally unicellular, but more usually have hyphae forming mycelium. Hyphae septate, with 1 (rarely more) perforations of septa. Cells uninucleate or multinucleate. Non-sexual reproduction by fission, budding, fragmentation or conidia which may be borne loosely or in non-sexual fruit bodies. In sexual reproduction, meiospores produced called ascospores, formed *inside* sac-like structures, each called an ascus, which may be naked, or more typically are assembled in fruit bodies called ascocarps. A large group of fungi including yeasts, some moulds and some large fungi such as morels, cup fungi, saddle fungi and truffles.

The classes of Ascomycotina

The classification of Ascomycotina is variable, depending on authority. For some time, five groups have been recognized, here given class status:

1. Hemiascomycetes (Protoascomycetes) in which the asci are naked.
2. Plectomycetes in which the asci are enclosed in a closed body called a cleistothecium.
3. Pyrenomycetes in which the asci develop inside a flask-shaped body called a perithecium, so are known as flask fungi.
4. Discomycetes in which the asci develop inside a cup- or disc-shaped body called an apothecium, and so are known as disc or cup fungi, although the name cup fungi is also used particularly for an order of Discomycetes, the Pezizales.
5. Loculomycetes in which the asci lie in cavities (loculi) surrounded by a two-layered wall.

As more fungi were investigated properly, these groups became unsatisfactory. A new class, Laboulbeniomycetes, was added to include the Laboulbeniales which was formerly in the Pyrenomycetes, and some orders of fungi were found to be so variable that they could be placed in the Pyrenomycetes, but could with equal validity be transferred to the Plecto- or Discomycetes. So a simpler and less equivocal classification was devised, having three classes (or equivalent) of ascomycetes:

1. Hemiascomycetes (Protoascomycetes) in which the asci are naked.
2. Euascomycetes in which the ascus wall is single-layered; the group includes most members of the Plectomycetes, Pyrenomycetes, Discomycetes and some other groups.
3. Loculomycetes (Loculoascomycetes) in which the asci lie in cavities (loculi) surrounded by a 2-layered wall.

If the ascomycetes are given class status, these classes become subclasses, and Plecto-, Pyreno- and Discomycetes become common names within the subclasses.

A summary of the alternative systems is given in Table 5.3; the moveable orders are shown with arrows indicating the different positions in which they may be found.

Table 5.3: Alternative classifications of ascomycetes

I	II
Subdivision Ascomycotina	Class Ascomycetes
Class Hemiascomycetes	Subclass Hemiasco-mycetidae
Order Endomycetales	
Order Taphrinales	Orders as in I
Order Protomycetales	
[Euascomycetes]	Subclass Euascomycetidae
Class Plectomycetes	[Plectomycetes]
Order Ascosphaerales	
Order Eurotiales	Orders as in I
Order Erysiphales	
Class Pyrenomycetes	[Pyrenomycetes]
Order Hypocreales	
Order Clavicipitales	
Order Chaetomiales	
Order Diaporthales — Sphaeriales	Orders as in I
Order Xylariales	
Order Coronophorales	
Order Coryneliales	
Order Microascales	
Order Meliolales	
Order Onygenales	
Class Discomycetes	[Discomycetes]
Order Phacidiales	
Order Helotiales	
Order Ostropales	Orders as in I
Order Pezizales	
Order Tuberales	
Class Laboulbeniomycetes	
Order Laboulbeniales	Order Laboulbeniales
Class Loculoascomycetes	Subclass Loculoas-comycetidae
Order Myriangiales	
Order Dothideales	
Order Pseudosphaeriales	
Order Pleosporales	Orders as in I
Order Hysteriales	
Order Microthyriales	
Order Capnodiales	

The classification described here is that in the left-hand column, with references to the right-hand column where appropriate.

Class Hemiascomycetes

(Protoascomycetes)

A single cell or hypha develops directly into an ascus, so ascus wall is cell wall.

No ascocarp or ascogenous hyphae. Saprophytic or parasitic.

Order Endomycetales

(Saccharomycetales, Protoascales)

Zygote or single cell transforms directly into ascus. Mycelium may be absent; some forms unicellular. Mostly saprophytic, a few parasitic.
e.g. *Endomyces, Saccharomyces* (yeast).

Order Taphrinales

(Exoascales)

Asci produced from binucleate cells of hyphae and form layer at surface of host plant. Mycelium present. Parasitic on vascular plants.
e.g. *Taphrina (Exoascus)* causes some witches' brooms.

Order Protomycetales

Ascus compound and called a synascus, produced from thick-walled spore in host plant. Mycelium present. Plant parasites.
e.g. *Protomyces, Protomycopsis, Taphridium*.

[Euascomycetes]

Ascus wall single-layered.

Class Plectomycetes

Asci borne in cleistothecium.

Order Ascosphaerales

Asci grouped into balls within a sac-like structure.
1 genus: *Ascosphaera*.

Order Eurotiales

(Aspergillales, Plectascales)

Asci globose to broadly oval, borne at different levels in cleistothecia. Saprophytes, parasites and human pathogens.
e.g. *Aspergillus (Eurotium)*.

The Onygenales may be included in the Eurotiales or placed as a separate order in the Pyrenomycetes.

Order Erysiphales

(Perisporiales, powdery mildews)

Asci globose to broadly oval, borne in cleistothecia with appendages. Obligate parasites on flowering plants, causing diseases called powdery mildews.
e.g. *Erysiphe, Microsphaera, Uncinula*.

Class Pyrenomycetes

(flask fungi)
Asci borne in perithecia.

Order Hypocreales

Perithecia and stromata (where present) often brightly coloured, soft, fleshy or waxy when fresh. Asci borne in basal layer among apical paraphyses. Saprophytes and parasites, some with conspicuous fruit bodies.
e.g. *Gibberella, Nectria* (coral spot).

Order Clavicipitales

Perithecia immersed in a stroma which comes from a sclerotium. Asci with thick apex penetrated by central canal through which ascospores ejected. Parasitic in plants, some in insect larvae.
e.g. *Claviceps* (ergot), *Cordyceps*.

Order Sphaeriales

Perithecia dark or black and more or less hard and carbonaceous. Usually saprophytes but some parasites.

This order may be divided into the following three orders:

Order Chaetomiales

Perithecia superficial and have conspicuous hairs on surface. Asci evanescent. Mainly saprophytes.
e.g. *Chaetomium*.

Order Diaporthales

Perithecia immersed in plant tissue or stroma with ostioles protruding. Asci on stalks which gelatinize, freeing asci from attachment; no paraphyses. Plant parasites and saprophytes.

e.g. *Diaporthe, Endothia* (chestnut blight).

Order Xylariales

Perithecia with dark membranous or carbonaceous walls, with or without a stroma. Asci persistent and borne in basal layer among paraphyses which may gelatinize and disappear. A large group of mainly saprophytes, some with conspicuous fruit bodies.
e.g. *Neurospora, Sordaria, Xylaria.*

Order Coronophorales

Asci in ascostromata with irregular or round, but never funnel-shaped ostioles.
e.g. *Bertia, Coronophora, Nitschkia, Scortechinia.*

Order Coryneliales

Asci in ascostromata with funnel–shaped ostioles at maturity.
e.g. *Corynelia, Coryneliospora, Tripospora.*

Order Microascales

Asci borne at different levels in perithecia. Asci evanescent. Some serious plant parasites.
e.g. *Ceratocystis (C. ulmi* – Dutch elm disease, *C. fagaceum* – oak wilt), *Microascus.*
   This order may be placed in the Plectomycetes.

Order Meliolales

Perithecia have no appendages. Asci borne in basal layers in perithecia. Mostly tropical parasites on leaves and stems of vascular plants; mycelium dark and superficial, and bearing appendages.
e.g. *Amazonia, Appendiculella, Asteridiella, Irenopsis, Meliola.*
   This order may be placed in the Plectomycetes.

Order Onygenales

Asci evanescent and borne in a mazaedium, and liberate ascospores as powdery mass among the sterile threads.
e.g. *Onygena, Onygenopsis.*
   This order does not have perithecia or cleistothecia, and may be placed in

the Plectomycetes or included in the Eurotiales.

Class Discomycetes

(disc or cup fungi)

Asci borne in apothecia.

Order Phacidiales

Apothecium immersed in black stroma, upper covering of which splits in stellate or irregular fashion when ascospores mature. Saprophytes or plant parasites.

e.g. *Lophodermium, Rhytisma.*

Order Helotiales

Apothecium typically somewhat saucer-shaped and bears inoperculate asci exposed from early stage; apothecia dehisce by a pore. Saprophytes and plant parasites.

e.g. *Ascocorticium, Sclerotinia.*

Order Ostropales

Ascocarp a locule-like apothecium and inoperculate with long narrow asci and thread-like ascospores.

e.g. *Ostropa, Stictis, Vibrissea.*

This order may be placed in the Pyrenomycetes.

Order Pezizales

(cup fungi)

Apothecium bears hinged cap, often large, and dehisces by this lid. Saprophytes.

e.g. *Helvella* (false morel), *Morchella* (morel).

Order Tuberales

(truffles)

Apothecia borne below ground and somewhat modified, with asci that are globose, oval or club-shaped, and do not dehisce. Mostly subterranean saprophytes.

e.g. *Terfezia* (Italian truffle), *Tuber* (truffle).

Class Laboulbeniomycetes

Minute parasites of insects and arachnids, with mycelia represented only by haustoria and stalks, with perithecia, and going no deeper than chitin. Of uncertain affinities, but may be considered Pyrenomycetes as the ascocarp is a perithecium.

Order Laboulbeniales

Characteristics as of class.
e.g. *Ceratomyces, Laboulbenia, Stigmatomyces.*

Class Loculoascomycetes

Ascus wall 2-layered. Asci borne in ascostromata.

Order Myriangiales

Asci borne at random at various levels in globose stroma. Tropical and parasitic on plants and sometimes insects.
e.g. *Molleriella, Myriangium, Uleomyces.*

Order Dothideales

(Dothiorales)

Asci in groups in a locule, with substantial opening produced by breakdown of surface layer; pseudoparaphyses absent. Usually plant parasites.
e.g. *Didymella, Dothidea, Dothiora, Mycosphaerella.*

Order Pseudosphaeriales

Asci borne among pseudoparaphyses in fruit body which opens by a pore or canal. Usually plant parasites.
e.g. *Leptosphaerulina, Pseudoplea.*

   This order may be included in the Dothideales, which may also then be called Pseudosphaeriales.

The following four orders may be included in the Pseudosphaeriales, because all bear asci among pseudoparaphyses.

Order Pleosporales

Asci borne among pseudoparaphyses in basal layer. Parasites or saprophytes.
e.g. *Lophiostoma, Pleospora, Venturia.*

Order Hysteriales

Asci borne among pseudoparaphyses, with ascostromata boat-shaped and opening by longitudinal slits. Saprophytes on wood.
e.g. *Glonium, Hysterium, Lophium.*

Order Microthyriales

Asci borne among pseudoparaphyses, with ascostroma flattened, hemispherical, opening by pore or tear, but base lacking. Saprophytes and leaf parasites.
e.g. *Asterina, Brefeldiella, Leptopeltis, Microthyrium.*

Order Capnodiales

(sooty moulds)

Asci borne among pseudoparaphyses. Ascostroma small, spherical, opening by pore or dissolution of peripheral tissue. Saprophytes and parasites.
e.g. *Capnodium, Limacina.*

Subdivision Basidiomycotina

(basidiomycetes)

Parasitic on plants and insects, or saprophytic. Always have hyphae forming mycelium. Hyphae septate, usually with dolipore septa. Mycelia differentiate into 2 types, (a) primary mycelium of uninucleate cells, (b) secondary mycelium of dikaryotic cells, these often bearing clamp connections over septa. Non-sexual reproduction by fragmentation, oidea or conidia. In sexual reproduction meiospores produced called basidiospores, formed on *outside* of cell called basidium, which may be found singly on hypha or assembled into fruit bodies called basidiocarps. A large group of fungi, including rusts, smuts, jelly fungi, bracket fungi, mushrooms and toadstools, puffballs, stinkhorns and bird's nest fungi.

The classes of Basidiomycotina

The classification of Basidiomycotina is variable, depending on authority. The main alternative systems of classification are given in Table 5.4.

The main differences, apart from the status of the groups, are:

1. The position of the Teliomycetes. This is due to the slight differences in the definitions of the terms Phragmo- and Heterobasidiomycetidae, which are only approximate synonyms, but is too complex to discuss further here.
2. The importance of the Hymenomycetes and Gastromycetes. This is because some authorities consider the characteristics that separate these groups to be of major taxonomic value, while others do not.

Table 5.4: Alternative classifications of basidiomycetes

Subdivision Basidiomycotina	Class Basidiomycetes

Subdivision Basidiomycotina

Class Teliomycetes
(Hemibasidiomycetes,
Promycetes)
    Order Uredinales
    Order Ustilaginales
Class Hymenomycetes
  Subclass
  Phragmobasidiomycetidae
  (≃Heterobasidiomycetidae)
    Order Tremallales
    Order Auriculariales
    Order Septobasidiales

  Subclass
  Holobasidiomycetidae
  (≃Homobasidiomycetidae)
    Order Exobasidiales
    Order Brachybasidiales
    Order Dacrymycetales
    Order Tulasnellales
    Order Polyporales
    Order Agaricales
Class Gastromycetes
    Order Hymenogastrales
    Order Lycoperdales ⎤Lycoperd-
    Order Sclerodermatales ⎦ales
    Order Sphaerobolales
    Order Nidulariales ⎤Nidulari-
    Order Phallales ⎦ales

Class Basidiomycetes
  Subclass Heterobasidiomycetidae
  (≃ Phragmobasidiomycetidae)

    Order Uredinales ⎤
    Order Ustilaginales ⎦ Teliomycetes

    Order Tremellales ⎤
    Order Auriculariales ⎟ Tremell-
    Order Dacrymycetales ⎟ ales
    Order Septobasidiales ⎦

  Subclass
  Homobasidiomycetidae
  (≃Holobasidiomycetidae)

    Order Polyporales ⎤ Hymenomycetes
    Order Agaricales ⎦

    Order Hymenogastrales
    Order Lycoperdales ⎤Lycoperd-
    Order Sclerodermatales ⎦ales
    Order Sphaerobolales ⎤Nidulari-
    Order Nidulariales ⎦ales
    Order Phallales ⎤ Gastromycetes

3. The position of the Dacrymycetales. This order may be included in the Tremellales because it has a gelatinous fruit body, or placed in the Holobasidiomycetidae because of the structure of the basidium.
4. The Tulasnellales was formerly included in the Tremellales.
5. The Brachybasidiales was formerly included in the Exobasidiales.

The classification described here is mainly that of the left-hand column, with reference to the right-hand column where appropriate.

Class Teliomycetes

(Hemibasidiomycetes, Promycetes)

Basidiocarp absent; basidial apparatus consists of a teliospore (teleutospore) which germinates into a promycelium on which basidiospores are borne.

Order Uredinales

(rust fungi)

Obligate parasites of vascular plants. Uredospores red, at least at first, giving the name rusts. Life cycle often involves 2 hosts. Life cycle complex and involves teliospores, basidiospores, pycnidiospores, aecidiospores and uredospores. Teliospores terminal. Basidiospores stalked and forcibly discharged. Many cause important plant diseases.

e.g. *Cronartium, Phragmidium, Puccinia, Uromyces.*

Order Ustilaginales

(smut or bunt fungi)

Not obligate parasites, and can be grown in laboratory cultures. Called smuts because teliospores black and dusty. Life cycle does not involve 2 hosts. Life cycle simpler than in rusts, and involves only teliospores and basidiospores, and sometimes conidia. Teliospores intercalary. Basidiospores sessile and not forcible discharged. Many cause important plant diseases.

e.g. *Tilletia, Ustilago.*

Class Hymenomycetes

Basidiocarp well developed and typically hymenium is exposed at maturity. Basidiospores forcibly discharged.

Subclass Phragmobasidiomycetidae

(approx. Heterobasidiomycetidae)

Basidia called phragmobasidia, vertically and/or transversely septate by primary septa (if Dacrymycetales included here basidia may also be deeply divided, i.e. Y-shaped); basidia may be described as being more than 1 cell deep.

Order Tremellales

(in the wide sense)

Basidiocarps well formed and gelatinous, but form inconspicuous horny crusts when dry.

   This order is placed in the Hemibasidiomycetes by some authorities.

This order may be divided into three or four orders:

Order Tremellales

(in the narrow sense)

Basidia divided longitudinally into 4, each bearing a basidiospore. These

basidia are also described as cruciate, i.e. cross-shaped, because the 2 longitudinal walls are perpendicular to each other.
e.g. *Hyaloria, Sirobasidium, Tremella.*

Order Auriculariales

Basidia divided transversely into 4, each bearing a basidiospore.
e.g. *Auricularia* (Jew's ear), *Helicobasidium.*

Order Dacrymycetales

Basidia divided into 2 (tuning fork basidia), each bearing a basidiospore.
e.g. *Dacrymyces, Ditiola.*

    The Dacrymycetales may be placed in the Holobasidiomycetidae, because its basidia are not divided by septa (as Phragmobasidiomycetidae are), or included here because its basidiocarp is gelatinous, and its basidia are divided, although not by septa (as Heterobasidiomycetidae are).

Order Septobasidiales

Basidiocarp not gelatinous, but basidia similar to Auriculariales. Development of basidia ceases for a time, then is resumed. Parasitic on, or symbiotic with, insects.
e.g. *Septobasidium.*

    Since the basidium is like that of the Auriculariales, this order may be included in the Auriculariales, or with them in the Tremellales.

Subclass Holobasidiomycetidae

(approx. Homobasidiomycetidae)

Basidia simple, non-septate, cylindrical or club-shaped, called holobasidia; basidia may be described as single-celled.

Order Exobasidiales

No basidiocarp; basidia 4-spored and produced in a layer on surface of leaf of parasitized vascular plant.
e.g. *Exobasidium, Kordyana.*

Order Brachybasidiales

No basidiocarp; basidia 2-spored and emerge through stomata of leaf of parasitized vascular plant.
1 genus: *Brachybasidium.*

    These two orders may be included together in the Exobasidiales since they are parasites of vascular plants, produce no basidiocarps, and form layers of spores on leaves.

Order Dacrymycetales

This order may be placed here rather than with the Tremellales, on the grounds that, although basidia are divided, they are not divided by septa.

For details and examples, see Tremellales.

Order Tulasnellales

Basidiocarps dry to gelatinous, but basidia not divided by septa, which places them here. Saprophytic and parasitic.

e.g. *Ceratobasidium, Tulasnella*.

This order may be included in the Tremellales.

Order Polyporales

(Aphyllophorales, Poriales)

Hymenium exposed for whole time and is developed at early stage. As fruit body enlarges, it extends and continues to develop gradually. Basidia borne in various ways, e.g. tubes, pores or grooves, but never gills. Most saprophytic.

e.g. *Cantharellus* (chanterelle – which looks as though basidia are on gills, but first 2 points above place it here), *Clavaria, Fomes* (tinder), *Hydnum* (wood hedgehog), *Polyporus* (Dryad's saddle).

Order Agaricales

Hymenium not exposed when fruit body very young, but enclosed in cavity, as in button mushroom. Whole hymenium developed at same time, not gradually. Hymenium borne on gills, or more rarely tubes, covering underside of stalked cap. Some parasites (e.g. honey fungus), some form mycorrhiza, most are saprophytic.

e.g. *Agaricus* (mushroom), *Amanita*, etc., the familiar gill toadstools, and *Boletus* (although hymenium lines tubes instead of gills, but placed here on the first 2 points above).

In many classificiations, only the last two orders are considered to be Hymenomycetes.

Class Gastromycetes

(Gasteromycetes)

Basidiocarp well developed and typically hymenium enclosed at maturity. Basidiospores not forcibly ejected. Basidia enclosed in basidiocarp whose outer layer is called a peridium and which encloses an inner layer called the gleba, on which basidia and basidiospores develop.

▮▮▮ Order Hymenogastrales

(Gastrohymeniales)

Basidiocarps underground, or if on surface lie buried in humus, and remain closed, the gleba being fleshy or waxy at maturity and disintegrating into a slimy mass containing the spores.
e.g. *Hymenogaster, Melanogaster.*

▮▮▮ Order Lycoperdales

(puffballs)

Peridium soft, gleba dry and powdery at maturity and consists of small pale spores with capillitium. Hymenium present in early stages.
e.g. *Calvatia, Lycoperdum* (both called puffballs), *Geaster* or *Geastrum* (earth star).

▮▮▮ Order Sclerodermatales

(earth balls)

Resemble Lycoperdales, but peridium hard, encloses powdery gleba, and spores dark with some capillitium. Hymenium absent in early stages.
e.g. *Scleroderma.*

These two orders may be combined together in the order Lycoperdales, on the grounds that the presence or absence of hymenium is difficult to see, and in both the gleba is dry and powdery.

▮▮▮ Order Nidulariales

(bird's nest fungi)

Gleba separates into spheres which become thick-walled and hard and are called peridioles, 'eggs', while the peridium is a cup-like or goblet-like 'nest', the whole resembling a miniature bird's nest, 6 to 12 mm in diameter, and containing about 3 'eggs'.
e.g. *Crucibulum, Cyathus, Sphaerobolus* (fungal artillery).
Sphaerobolus may be removed to its own order, **Sphaerobolales**, on the grounds that its 'eggs' are forcibly discharged (hence its common name).

▮▮▮ Order Phallales

(stinkhorns)

Gleba slimy and foetid at maturity, and exposed on elongated or net-shaped structures, breaking through peridium which is left as volva-like structure at base.
e.g. *Clathrus* (lattice fungus), *Dictyophora, Mutinus* (dog's stinkhorn), *Phallus* (stinkhorn).

Subdivision Deuteromycotina

(deuteromycetes, Fungi Imperfecti)

Artificial group, comprising those fungi which are only known to reproduce non-sexually, or not at all. Many responsible for diseases. Since classification of most classes involves characteristics concerned with sexual reproduction, these cannot be assigned to any other group, although the non-sexual stages resemble those of asco- or basidiomycetes.

The classification of Fungi Imperfecti is variable depending on the status given to the groups. The alternative systems are shown in Table 5.5. The system described here is mainly that of the left-hand column, with references to the right-hand column where appropriate. To be completely correct, the terms form class and form order should be used rather than class and order, since relationships involving breeding cannot exist in this group.

Table 5.5: Alternative classifications of Fungi Imperfecti

Subdivision Deuteromycotina	Class Deuteromycetes
Class Blastomycetes	
Order Cryptococcales	
Order Sporobolomycetales	Order Moniliales
Class Hyphomycetes	= Blastomycetes + Hyphomycetes
Order Hyphomycetales	(excluding Agonomycetales)
Order Stilbellales	
Order Tuberculariales	
Order Agonomycetales	Order Mycelia Sterila
Class Coelomycetes	
Order Sphaeropsidales	Order Sphaeropsidales
Order Melanconiales	Order Melanconiales

Class Blastomycetes

Consists of yeast-like forms.

Order Cryptococcales

Reproduction by budding; spores not forcibly discharged (suggesting relationship with ascomycetes).
e.g. *Cryptococcus, Rhodotorula.*

Order Sporobolomycetales

Reproduction by budding; spores forcibly discharged (suggesting relationship with basidiomycetes).
e.g. *Bullera, Sporobolomyces.*

Class Hyphomycetes

Consists of mycelial forms; sterile, or bear conidia on hyphae, not in fruit bodies.

Order Hyphomycetales

Conidia present; conidiophores not organized as synnemata or sporodochia.
e.g. *Dematium, Monilia, Penicillium.*

Order Stilbellales

Conidia present; conidiophores organized as synnemata.
e.g. *Stilbella, Stilbellula.*

Order Tuberculariales

Conidia present; conidiophores organized as sporodochia.
e.g. *Tubercularia.*

All these groups of Fungi Imperfecti were formerly placed in the order **Moniliales,** whose characteristics were that conidia were borne on conidiophores, never in pycnidia or acervuli.

Order Agonomycetales

(Mycelia Sterila)
Spores usually absent.
e.g. *Papulospora, Rhizoctonia, Sclerotium.*

Class Coelomycetes

Consists of mycelial forms; conidia borne in pycnidia or acervuli.

Order Sphaeropsidales

Conidia borne in pycnidia.
e.g. *Septoria, Sphaeropsis.*

Order Melanconiales

Conidia borne on acervuli, or in layers on stromata.
e.g. *Colletotrichum, Gloeosporium.*

6 Lichens

Division Lichenes

(lichens)

Composite plants, associations of algae and fungi in relationship usually regarded as symbiotic, sometimes said to be parasitic by fungus. Alga is of Cyanophyta or Chlorophyta; fungus is ascomycete or basidiomycete. Alga can live freely without fungus, but fungus cannot survive without alga; but living with fungus enables alga to colonize dryer habitat. Only fungus reproduces sexually, but small bodies called soredia, consisting of a few algal cells and some fungal mycelium, can reproduce non-sexually. Very widespread early colonizers of bare habitats.

Lichens are usually regarded as a division in their own right, although they have been considered a subdivision of fungi. Since only the fungal portion can reproduce, they are sometimes classified into the classes Ascomycetes and Basidiomycetes, but these groups are also known as Ascolichenes and Basidiolichenes, the system adopted here.

Class Ascolichenes

(ascolichens, Ascomycetes)
Fruit bodies contain asci.

Subclass Ascomycetidae

Asci unitunicate (with 1 wall), regularly arranged in hymenium with unbranched paraphyses.

Order Lecanorales

Fruit bodies apothecia.
e.g. *Cladonia, Lecanora, Parmelia, Peltigera, Physcia, Usnea, Xanthoria.*

Order Sphaeriales

Fruit bodies perithecia.
e.g. *Pyrenula, Strigula, Verrucaria.*

Order Caliciales

Fruit bodies mazaedia.
e.g. *Calicium, Cyphelium, Sphaerophorus.*

Subclass Loculoascomycetidae

Asci bitunicate (with 2 walls), regularly or irregularly arranged in a pseudothecium (ascostroma) with branched paraphyses.

Order Myriangiales

Asci irregularly distributed, pseudothecia poorly differentiated.
e.g. *Arthonia, Mycoporellum.*

Order Pleosporales

Asci fairly regularly arranged in stroma; pseudothecia well differentiated, resembling perithecia.
e.g. *Leptorhaphis, Melanotheca.*

Order Hysteriales

Pseudothecia well differentiated, round and resembling apothecia, or irregular.
e.g. *Lecanactis, Opegrapha, Roccella.*

Former classification of ascolichens

Formerly the Ascolichenes were divided into two subclasses, the Pyrenocarpeae and Gymnocarpeae, whose classification was as follows:

Subclass Pyrenocarpeae

(a small group)

Fruit body a perithecium.
e.g. *Dermatocarpon, Verrucaria.*

Subclass Gymnocarpeae

(a larger group)

Fruit body an apothecium, but often with thick walls. The orders were rather artificial and distinguished by one characteristic only, so are here separated by a key:

1. Contain blue-green algae	Cyanophilales e.g. *Collema, Ephebe, Lobaria, Peltigera*
Contain green algae	2
2. Spores generally thick-walled and 2-celled	Caloplacales e.g. *Caloplaca, Xanthoria*
Spores generally thin-walled and 1-celled	3
3. Apothecia elongated, resembling runic inscriptions	Graphidales e.g. *Graphis*
Apothecia not as above	4
4. Asci disintegrate so spores form loose mass with paraphyses which continue to elongate	Caliciales e.g. *Calicium, Sphaerophorus*
Asci not as above	5
5. Margins of apothecia free of algae .	Lecideales e.g. *Cladonia, Lecidea, Rhizocarpum*
Green algae continue to margins of apothecia	
	Lecanorales e.g. *Cetraria, Lecanora, Parmelia, Usnea*

Class Basidiolichenes

(basidiolichens, Basidiomycetes)

Fruit bodies contain basidia.

e.g. *Botrydina, Cora, Dictyonema, Herpothallon, Omphalina.*

There are also some lichens whose fruit bodies are unknown, which are assigned to the fungal group Fungi Imperfecti, or are known as **Lichenes Imperfecti**, e.g. *Lepraria, Lichenothrix.*

7 Bryophyta

(bryophytes)

Photosynthetic, non-vascular plants in which female sex organ is archegonium, and which show clear alternation of generations. Gametophyte generation dominant, conspicuous and independent. Gametophyte thalloid or organized into leaf-like structures (phyllids), not true leaves. Sporophyte generation consists of stalk and capsule (together called sporogonium) and is as conspicuous as gametophyte, but is dependent on it for the whole of its lifetime and cannot live an independent existence, although photosynthesis occurs in lower part of capsule of mosses. Ovum remains in archegonium and spermatozoid swims to it by chemotaxis to effect fertilization. Spores dispersed by mechanism ensuring dispersal in dry weather only. No vascular tissue in either generation. No true roots, stems or leaves in either generation; phyllids do not possess cuticle or stomata; root-like structures, rhizoids, are either unicellular or filaments of colourless cells.

Classes of bryophytes

Systems of classification usually consider that there are either two or three classes of bryophytes. The two classes are the Bryopsida (formerly called Musci), the mosses, and the Hepaticopsida (formerly called Hepaticae), the liverworts. There is a third group, the hornworts or horned liverworts, which may be considered to be an order of liverworts (Anthocerotales), or raised to the level of a subclass, Anthocerotidae, with the true liverworts as the Hepaticidae, within the class Hepaticopsida. The hornworts are very different from other liverworts and now are more usually raised to class status as Anthocerotopsida, making three classes, Bryopsida, Hepaticopsida and Anthocerotopsida. This is the system adopted here.

Summary classification of the division Bryophyta

Class Bryopsida (Musci, mosses)

  Subclass Sphagnidae (peat or bog mosses)

Order Sphagnales

Subclass Andreaeidae (granite mosses)

Order Andreaeales

Subclass Bryidae (true mosses)

Cohort Eubryiidae

Order Archidiales

Order Dicranales

Order Fissidentales

Order Encalyptales

Order Pottiales

Order Grimmiales

Order Funariales

Order Eubryales (Bryales)

Order Orthotrichales ⎤
Order Isobryales   ⎦ Isobryales

Order Hookeriales

Order Hypnobryales ⎤
Order Thuidiales   ⎦ Hypnobryales

Order Schistostegales

Order Tetraphidales (position uncertain)

Cohort Buxbaumiidae

Order Buxbaumiales ⎤
Order Diphysciales   ⎦ Buxbaumiales

Cohort Polytrichiidae

Order Polytrichales

Order Dawsoniales

Class Hepaticopsida (true liverworts)

Subclass Jungermanniae

Order Takakiales

Order Calobryales

Order Jungermanniales ⎤
Order Metzgeriales   ⎦ Jungermanniales

Subclass Marchantiae

Order Sphaerocarpales

Order Monocleales

Order Marchantiales

Class Anthocerotopsida (horned liverworts or hornworts)

Order Anthocerotales

Some authorities raise the mosses and liverworts to division status as Bryophyta (mosses) and Hepaticophyta (liverworts), but this is confusing and is not recommended.

Class Bryopsida

(Musci, mosses)

Mature gametophyte never thalloid. Phyllids not arranged in clear rows. Gametophyte erect or prostrate. Rhizoids multicellular. Protonema usually extensive and filamentous, sometimes thalloid. Sporophyte with spherical or cylindrical capsule. Seta elongates gradually. Capsule has tough epidermis and opens by operculum or by 4 slits. Capsule usually has sterile columella in centre. Dehiscence in dry weather and often controlled by hygroscopic peristome teeth. Lower part of capsule may be photosynthetic with stomata.

The Bryopsida is divided into three groups; these may be considered to be of the level of subclasses and known as Sphagnidae, Andreaeidae and Bryidae (as here) or may be reduced to the level of orders as Sphagnales, Andreaeales and Bryales, or even raised to the level of classes as Sphagnopsida, Andreaeopsida, Bryopsida in those classifications where mosses and liverworts are treated as separate divisions.

Subclass Sphagnidae

(peat or bog mosses)

Protonema usually thalloid. Gametophyte at maturity usually erect, much branched, with many phyllids which are absorbent because they possess large empty cells with pores in walls. No seta; sporophyte borne on pseudopodium. Capsule spherical with operculum but no peristome. Spore-bearing tissue thimble-shaped, with dome-like columella. Stomata absent. Occur in bogs, acid, wet places. Formerly called Sphagnobrya.

Order Sphagnales

Characteristics as of subclass.
1 genus: *Sphagnum*, contributes to formation of sphagnum peat.

Subclass Andreaeidae

(granite mosses)

Protonema usually thalloid. Gametophyte usually erect, small, dark, with many compactly arranged phyllids. No seta, sporophyte borne on short pseudopodium. Capsule cylindrical, without operculum, and opens by 4 longitudinal slits. Spore-bearing tissue shaped like a long thimble over sterile, finger-shaped columella. Stomata absent. Usually found on exposed non-calcareous rock, occasionally on sand in high altitude stream beds. Formerly called Andreaeobrya.

Order Andreaeales

Characteristics as of subclass.
2 genera *Andreaea* and *Neuroloma*.

Subclass Bryidae

(true mosses)

Protonema filamentous. Gametophyte erect or prostrate, usually with many phyllids, variously coloured. Seta present in most species, and pseudopodium always absent. Capsule cylindrical to spherical, usually with deciduous operculum and peristome. Spore-bearing tissue shaped like hollow cylinder around central columella. Stomata often present in basal half of capsule. Occur in various substrata in many environments. Formerly called Eubrya.

The Bryidae may be divided into three groups, which may be considered as three orders, the Eubryales, Buxbaumiales and Polytrichales. They are very different from one another and are now usually given a status higher than that of an order. Here this status is called a cohort, and the groups are called Eubryiidae, Buxbaumiidae and Polytrichiidae. A cohort is not now a recognized ICBN term, and in some classifications these groups are raised to subclass status, raising the groups here considered as subclasses to class status.

Cohort Eubryiidae

Peristome single or double; outer teeth appear barred.

Order Archidiales

Capsule unstalked, globose, without columella, lid or peristome.
e.g. *Archidium*.

Order Dicranales

Acrocarpous. Peristome single with 16 bifid peristome teeth. Capsule ovoid or cylindrical.
e.g. *Ceratodon, Dicranum*.

Order Fissidentales

Acrocarpous. Peristome as in Dicranales. Capsule variable. 2 very clear rows of phyllids, with wing from back of midrib, and nerve prolonged into wing.
e.g. *Fissidens*.

Order Encalyptales

Tufted growth. Peristome variable, well developed and double, or single and rudimentary. Capsule cylindrical.
e.g. *Encalypta*.

Order Pottiales

Acrocarpous, usually small. Peristome single, often spirally twisted; teeth divided into segments or absent. Capsule ovoid to cylindrical. Cells of phyllids and outer side of peristome teeth generally papillose.
e.g. *Pottia, Tortula*.

Order Grimmiales

Acrocarpous. Peristome single. Capsule spherical to cylindrical. Each phyllid has long colourless hair as continuation of midrib. Often tufted and growing on rock.
e.g. *Grimmia, Rhacomitrium*.

Order Funariales

Acrocarpous. Peristome usually double; inner teeth opposite outer teeth, inner lacking basal membrane and cilia. Capsule usually pendulous and never cylindrical. Usually terrestrial.
e.g. *Funaria, Physcomitrium*.

Order Eubryales

(Bryales)

Acrocarpous. Peristome double, inner well developed and usually with cilia.
e.g. *Bryum, Mnium,*

Order Orthotrichales

Acrocarpous, but may appear pleurocarpous. Peristome double, outer of 16 teeth, inner thin or absent, or capsule without peristome.
e.g. *Amphidium, Orthotrichum*.
 This order is sometimes included in the Isobryales.

Order Isobryales

Pleurocarpous. Peristome double. Calyptra mitre- or hood-shaped.
e.g. *Climacium, Thamnobryum.*

Order Hookeriales

Pleurocarpous. Peristome double. Calyptra conical, often fringed. Phyllids frequently asymmetrical. Mainly tropical.
e.g. *Daltonia, Hookeria.*

Order Hypnobryales

Pleurocarpous; primary stems variously branched. Peristome double. Capsule inclined to horizontal, rarely erect. A very large order.
e.g. *Amblystegium, Brachythecium, Hypnum, Plagiothecium.*

Order Thuidiales

Pleurocarpous; primary stems creeping, secondary stems ascending and regularly or irregularly branched. Peristome double. Capsule erect or inclined.
e.g. *Heterocladium, Myurella, Thuidium.*
    This order may be included in the Hypnobryales.

Order Schistostegales

Luminous due to light-refractive properties of persistent thalloid protonema.
1 genus and species: *Schistostega pennata*, the luminous moss.

Order Tetraphidales

Of uncertain position, because its 4 peristome teeth are intermediate between Eubryiidae and Polytrichiidae.
e.g. *Tetraphis.*

Cohort Buxbaumiidae

Peristome single or double; outer peristome of 1 to 4 rows of teeth, which appear barred; inner peristome membranous, folded longitudinally. Capsule very large and asymmetrical.

Order Buxbaumiales

Outer peristome of 1 to 4 rows of teeth; inner membranous with 32 longitudinal folds.

1 genus: *Buxbaumia*.

Order Diphysciales

Outer peristome absent or rudimentary; inner membranous with 16 longitudinal folds.

1 genus: *Diphyscium*.

These two orders may be combined in the same order, Buxbaumiales, which then has the characteristics of the cohort.

Cohort Polytrichiidae

Peristome teeth usually 32 or 64, made of horseshoe-shaped cells; do not appear barred but solid structure.

Order Polytrichales

Acrocarpous. Peristome of short teeth, joining at their tip to a flat membranous structure called an epiphragm across mouth of capsule, with spores dispersed through holes between teeth.

e.g. *Atrichum, Oligotrichum, Polytrichum*.

Order Dawsoniales

Large acrocarpous mosses. Peristome represented by bunch of bristles.

1 genus: *Dawsonia* (of Australasia).

Class Hepaticopsida

(Hepaticae, true liverworts)

Gametophyte leafy or thalloid; where leafy, very dorsiventral. In leafy forms, 2 lateral rows of 'leaves', phyllids, and sometimes a smaller ventral row of 'leaves', amphigastria. Gametophyte usually prostrate, with several chloroplasts in each cell. Rhizoids unicellular. Protonema less extensive than in mosses. Sporophyte with apical growth and produces globular capsules. Seta elongates quickly. Capsule has undifferentiated walls and opens by splitting longitudinally (or, rarely, by erosion). Capsule without columella. Dehiscence in dry weather and often controlled by hygroscopic elaters between cells. Capsule not photosynthetic and does not have stomata. Elaters unicellular.

The subclasses of Hepaticopsida do not have ICBN endings.

Subclass Jungermanniae

Sporophyte well developed; capsule on long seta, and usually opening by 4 valves. Gametophyte with little tissue specialization.

Order Takakiales

Gametophyte leafy with cylindrical phyllids, almost in whorls, resembling a miniature stonewort; archegonia without perianth of bracts and contain chlorophyll; antheridia, sporophyte and rhizoids unknown.
1 genus: *Takakia.*
   A little-known order which may be included in the Calobryales.

Order Calobryales

Gametophyte leafy with radial symmetry and resembling mosses, with 3 identical rows of phyllids; upright branches arise from buried, rhizome-like stems. Rhizoids absent. Archegonia not surrounded by perianth of bracts; archegonia have 4 vertical rows of neck cells, less than other groups.
1 genus: *Haplomitrium* (including *Calobryum*).

Order Jungermanniales

(in the wide sense)

Leafy or thalloid. Rhizoids of 1 kind. Antheridia and archegonia enclosed by perianth of bracts. Capsule on elongated, fragile seta. Air chambers absent. Elaters always present.
   This order is now usually separated into two orders, the Jungermanniales and Metzgeriales:

Order Jungermanniales

(in the narrow sense)

Leafy liverworts, with phyllids divided into upper and lower lobes, and with amphigastria underneath.
e.g. *Calypogeia, Lepidozia, Lophocolea.*

Order Metzgeriales

Usually thalloid; if phyllids develop they are never lobed, and no amphigastria.
e.g. *Metzgeria, Pellia.*

Subclass Marchantiae

Sporophyte reduced; seta short or absent, capsule opening by various mechanisms, but seldom by 4 valves. Gametophyte with much tissue specialization.

Order Sphaerocarpales

Gametophyte small, delicate, with simple thallus, resembling fern prothalli. Rhizoids of 1 kind. Antheridia and archegonia enclosed in pear-shaped perianth with apical opening. Air chambers absent. Elaters absent.
e.g. *Riella, Sphaerocarpos (Sphaerocarpus)*.

Order Monocleales

Formerly included in Marchantiales, but: sporophyte less reduced; no air chambers; seta long.
1 genus: *Monoclea*.

Order Marchantiales

Gametophyte consists of dorsiventral complex thallus with internal differentiation. Rhizoids of 2 kinds. Archegonia and antheridia not surrounded by perianth, but often united on receptacle on long stalk. Air chambers present. Sporophyte reduced. Elaters usually present.
e.g. *Concephalum, Marchantia, Riccia*.

Class Anthocerotopsida

(Anthocerotae, horned liverworts, hornworts)
Differ from true liverworts in that:
Gametophyte always thalloid. Gametophyte with single chloroplast in each cell. Sporophyte with basal growth and produces a needle-shaped capsule with columella and stomata. Elaters multicellular (pseudoelaters) or absent.

Order Anthocerotales

Characteristics as of class.
e.g. *Anthoceros, Dendroceros, Notothylas*.

8 Pteridophyta

The Pteridophyta may be considered to be a division, or it may be raised to higher status, and its classes themselves considered as divisions. Here it is given division status.

Division Pteridophyta

(pteridophytes)

Photosynthetic vascular plants in which female sex organ is an archegonium, and which show clear alternation of generations. Sporophyte generation dominant and conspicuous, but dependent on gametophyte generation early in life. Gametophyte is free-living generation, but very reduced, called a prothallus, and bears antheridia and archegonia on same or different prothalli; prothallus consists of thin plates of cells, lacks a cuticle, and is sometimes subterranean or absent. Sporophyte generation consists of true roots, stems and leaves, all with vascular tissue, and leaves possess cuticle and stomata (except in some fossil forms and primitive modern members). Ovum remains in archegonium, and spermatozoid swims to it by chemotaxis to effect fertilization. Spores dispersed by mechanism ensuring dispersal in dry weather; plants homosporous or heterosporous; in heterosporous types large megaspores give rise to female prothalli bearing archegonia, and small microspores give rise to male prothalli bearing antheridia. Vascular tissue usually confined to sporophyte and arranged in various kinds of steles. Xylem usually of tracheids only, vessels rare. Leaves megaphyllous or microphyllous. In living pteridophytes, sporangia usually borne on leaves, or on stems associated with them or on modified leaf-like structures associated with them (sporophylls); but Devonian forms had no leaves and bore sporangia on their stems.

Classes of pteridophytes

Four or five classes are recognized. These are the Psilophytopsida and Psilotopsida (which may be combined together to make the class Psilopsida), the Lycopsida, Sphenopsida and Filicopsida. These four classes may be raised to division level and are then known as Psilophyta, Lycophyta,

Table 8.1

Division or	Class		Common name
Psilophyta	Psilopsida	⌈ Psilophytopsida ⌊ Psilotopsida	whisk ferns
Lycophyta or Lepidophyta or Lycopodiophyta	Lycopsida or Lycopodiatae or Lycopodineae		club mosses
Sphenophyta or Arthrophyta or Calamophyta or Equisetophyta	Sphenopsida or Articulatae (Articulae) or Equisetatae or Equisetineae		horsetails
Filicophyta or Pterophyta or Polypodiophyta	Filicopsida or Pteropsida or Polypodiopsida or Filicatae or Filicineae		ferns

Sphenophyta and Filicophyta. Each class and division have other names as shown in Table 8.1.

The term Pteropsida can also be used for a group consisting of the ferns, gymnosperms and flowering plants on the supposition that all three groups are closely related.

Here the groups are considered as five classes of pteridophytes: Psilophytopsida, Psilotopsida, Lycopsida, Sphenopsida, Filicopsida.

Summary classification of the division Pteridophyta

Class Psilophytopsida

    Order Psilophytales

Class Psilotopsida

    Order Psilotales

Class Lycopsida

    Order Protolepidodendrales

    Order Lycopodiales

    Order Lepidodendrales

    Order Pleuromeiales

    Order Isoetales

    Order Selaginellales

Class Sphenopsida

    Order Hyeniales (Protoarticulatae)

    Order Pseudoborniales

    Order Sphenophyllales

    Order Calamitales

    Order Equisetales

    Order Noeggerathiales (position uncertain)

Class Filicopsida

  A. Primofilices

    Order Protopteridales*

    Order Cladoxylales

    Order Coenopteridales

    Order Archaeopteridales*

  B. Eusporangiatae

    Order Marattiales

    Order Ophioglossales

  C. Osmundidae

    Order Osmundales

  D. Leptosporangiatae (Filices)

    Order Filicales

   Hydropterides

    Order Marsileales

    Order Salviniales

*Also considered to be progymnosperms.
A–D may be regarded as subclasses, although they do not have consistent ICBN subclass endings.

Class Psilophytopsida

Extinct primitive plants, with sporophyte whose rhizomes and aerial branches are more or less dichotomous; rhizoids instead of roots for absorption. Stems with small appendages, spirally arranged, or stems naked; only sporophyte known. Protostelic. No secondary thickening. Homosporous. Sporangia borne terminally or laterally on naked branches, not associated with leaves. Sporangia thick-walled. Spermatozoids and gametophyte generation unknown. Abundant in Devonian period.

Order Psilophytales

Characteristics as of class.
e.g. *Horneophyton, Psilophyton, Rhynia, Yarravia, Zosterophyllum. Asteroxylon* was formerly included here, but has now been transferred to the Lycopsida; *Pseudosporochnus* was placed here, and has now been transferred to the Cladoxylales.

Class Psilotopsida

Primitive living forms; have rootless sporophyte with aerial shoots arising from dichotomously branched rhizomes; rhizoids instead of roots for absorption. Leaf-like or scale-like lateral appendages spirally arranged; stem acts as photosynthetic organ. Protostelic. No secondary thickening. Homosporous. Sporangia borne on very short lateral branches. Sporangia thick-walled. Exosporic. Spermatozoids multiflagellate. Only living types known.

Order Psilotales

Characteristics as of class.
2 genera: *Psilotum, Tmesipteris.*

Class Lycopsida

(Lycopodiatae, lycopods, club mosses)

Sporophyte with roots, simple or dichotomously branched stems, and leaves on stems. Leaves microphyllous and sessile; rhizophores may be present. Protostelic, sometimes polystelic, occasionally polycyclic. Some with secondary thickening. Homosporous or heterosporous. Sporangia borne on or near a sporophyll. Sporangia thick-walled. Exosporic and endosporic. Spermatozoids biflagellate or multiflagellate. Tree lycopods appeared in Upper Devonian and survived until Lower Permian; smaller forms survive today; some (not trees) known from Cambrian and Silurian.

Order Protolepidodendrales

Extinct forms only. Homosporous. Dichotomously branched; may have arisen from horizontal rhizome. Probably no secondary thickening. No ligule. Not known whether exo- or endosporic. No information about number of flagella on sperm.
e.g. *Aldanophyton* [Cambrian], *Baragwanathia* [Silurian], *Drepanophycus* and *Protolepidodendron* [Devonian].

Order Lycopodiales

Living and extinct forms known. Homosporous. Show dichotomous branching but are not trees. Primary growth only. No ligule. Exosporic. Sperm biflagellate.

2 living genera: *Lycopodium, Phylloglossum*; and an extinct Carboniferous form, *Lycopodites*.

The term club moss may be used for the whole class Lycopsida, or just for the order Lycopodiales, or just for the genus *Lycopodium*.

Order Lepidodendrales

(lepidodendroids)

Only extinct Carboniferous forms known. Heterosporous. Tree lycopods. Some show secondary thickening as well as primary growth. Ligule present. Endosporic. No information about flagella on sperm. *Lepidocarpon* has megaspore retained in sporangium on sporophyte, and shed like a seed.

e.g. *Lepidocarpon, Lepidodendron, Lepidophloios, Sigillaria, Stigmaria. Pleuromeia* may also be included here or in an order of its own, Pleuromeiales; see below.

The retention of the megaspore in *Lepidocarpon* has been said to be a seed-like structure, and these forms were formerly thought to provide a link between the pteridophytes and the pteridosperms of the spermatophytes. This is now thought not to be so, the retained megaspore not being a true seed.

Order Pleuromeiales

Extinct Triassic forms. Heterosporous. Unbranched, with underground rhizome having 4 blunt lobes from which arise slender branching rootlets. Probably no secondary growth, but internal tissue not well preserved. Ligule present. No evidence whether exo- or endosporic, or of flagellation of sperm.

1 genus: *Pleuromeia*.

This genus is thought to be a link between the Isoetales and Lepidodendrales, so it is placed in its own linking order, or may be included in the Lepidodendrales, or sometimes in the Isoetales.

Order Isoetales

(quillworts)

Living and extinct forms known. Heterosporous. Linear leaves grow from short tuberous stem known as a 'corm'. Unusual secondary growth with anomalous cambium. Ligule present. Endosporic. Sperm multiflagellate. Includes living genera *Isoetes* and *Stylites*, and extinct Cretaceous genus

Isoetites, and sometimes *Nathorstiana* which may also be placed in the Pleuromeiales.

Order Selaginellales

Living and extinct forms known. Heterosporous. Leaves ranked in 2 layers on stem. Primary growth only. Ligule present. Endosporic. Sperm biflagellate. 1 living genus, *Selaginella*, and 1 extinct Carboniferous genus, *Selaginellites*.

Class Sphenopsida

(Equisetatae, Articulatae, horsetails)

Sporophyte with roots, and rhizomes from which emerge upright jointed stems and leaves. Leaves microphyllous, wedge-shaped in living forms, arranged in whorls at nodes with internodes in between; stems green, ribbed, photosynthetic. Protostelic. Some with secondary thickening. Homosporous or heterosporous. Sporangia borne in reflexed position (except in *Protohyenia*) on usually peltate sporangiophores arranged in whorls, usually on terminal cone. Sporangia with thick walls. Exosporic in living species. Spermatozoids multiflagellate. Horsetails appeared in Lower Devonian, and were important in late Palaeozoic and Mesozoic forests, but *Equisetum* is only genus surviving today.

Order Hyeniales

(Protoarticulatae)

Extinct. Shrub-like. Short, forked appendages in whorls. Fossils suggest that there may have been secondary thickening. Homosporous. Found in Lower to Mid Devonian rocks.
e.g. *Calamophyton, Hyenia, Protohyenia*.

Order Pseudoborniales

Extinct. Known from compressions, not petrified remains, and consist of jointed stems of various dimensions, bearing whorls of large deeply incised leaves, resembling those of Sphenophyllales. Sporangia found containing spores believed to be megaspores.
1 species: *Pseudobornia ursina*, found in Upper Devonian rocks of Bear Island.

Order Sphenophyllales

Extinct. Vine-like, prostrate or scrambling. Leaves small, wedge-shaped or forked and arranged in whorls of about 6. Secondary thickening present.

Mostly homosporous, occasionally heterosporous. Found from Upper Devonian to Lower Triassic times.

e.g. *Bowmanites, Cheirostrobus, Eviostachya, Sphenophyllum.*

Order Calamitales

Extinct. Of tree form. Leaves needle-like, wedge-shaped, strap-like or forked, in whorls of various number. Secondary thickening present. Homo- and heterosporous. Found from Upper Devonian to Permian times, reaching peak of development in Upper Carboniferous.

e.g.   *Asterocalamites,   Archaeocalamites,   Calamites,   Palaeostachya, Protocalamites.*

Order Equisetales

Living and extinct forms known. Herbaceous plants. Leaves scale-like, forming a whorl at nodes, and from which arise a whorl of branches. No secondary thickening. Homosporous, but spores develop into 2 kinds of prothalli; elaters for spore dispersal.

2 genera: 1 living genus, *Equisetum*, and 1 extinct, *Equisetites*, of Upper Carboniferous to Mesozoic.

Sometimes the name horsetail is restricted to this group. In some classifications, the Hyeniales and Calamitales are included in the Equisetales.

Order Noeggerathiales

This is an order of extinct plants of uncertain position, which can be placed with the Sphenopsida or ferns, and have also been considered as palms or cycads. Leaves cycad-like, cones unlike any other group of plants. Known from fossils of continental Europe from Permo-Carboniferous to Triassic.

2 genera: *Noeggerathia, Tingia.*

Class Filicopsida

(Filicatae, Pteropsida, Polypodiopsida, ferns)

Sporophyte with roots, stems (rhizomes) and leaves emerging from rhizomes. Leaves usually megaphyllous, known as fronds, and often divided into pinnae and pinnules. Protostelic, solenostelic, dictyostelic, sometimes polystelic, rarely polycyclic. Some with limited secondary thickening (rare). Homospor- ous or heterosporous. Sporangia borne on lower surface of foliage leaves or on axis or terminally; may be solitary or arranged in sori. Sporangia with thick or thin walls. Exosporic and endosporic. Spermatozoids multiflagellate. Ferns appeared in Devonian period, and were important in swamp forests of Upper

Carboniferous and Permian; just the remainder survive today.

There are many systems of classification of ferns. The system here uses terms with endings not recommended by the ICBN, but which are equivalent to subclasses.

A. Primofilices

These are a fossil group, probably ancestors of modern ferns, forming a link between them and their psilophyte ancestors. They have the following features:

Terminal sporangia. Usually homosporous, some heterosporous. Thick-walled sporangia (eusporangiate). Appeared in Devonian and became extinct in Lower Permian.

They may be grouped into two orders, the Cladoxylales and Coenopteridales, or into four orders, the Protopteridales, Cladoxylales, Coenopteridales and Archaeopteridales. Members of the Archaeopteridales and Protopteridales are also considered to be progymnosperms, but may be included here because, while ancestral to gymnosperms, they are at the level of pteridophytes in their method of reproduction. The groups are in chronological order below:

Order Protopteridales

Known from Lower to Mid Devonian. Resemble psilophytes. Homosporous. No annulus.

e.g. *Aneurophytum, Protopteridium.*

Order Cladoxyales

Known from Mid Devonian to Lower Carboniferous. Resemble psilophytes. Homosporous. No annulus.

e.g. *Cladoxylon, Pseudosporochnus* (formerly in Psilophytopsida).

Order Coenopteridales

Known from Upper Devonian to Lower Permian with maximum development in Lower Carboniferous. Fronds 3-dimensional in structure. Usually homosporous, but *Stauropteris* is heterosporous. Some have annulus.

e.g. *Ankyropteris, Botryopteris, Stauropteris.*

Order Archaeopteridales

Known from Upper Devonian to Lower Carboniferous. Fronds flattened and

sporangia borne on margins. Heterosporous. No annulus.
e.g. *Archaeopteris.*

B. Eusporangiatae

(eusporangiate ferns)

Sporangium eusporangiate, i.e. sporogenous tissue derived from inner daughter cell. Adjacent cells involved in formation of part of sporangium wall and stalk. Sporangium large and massive; wall several cells thick, annulus usually absent. Spore content high. Primitive.

Order Marattiales

(giant ferns)

Terrestrial. Leaves often pulvinate in living genera, and with fleshy stipules; leaves pinnately divided. Leaf not divided into sterile and fertile segments, but sporangia in sori or more or less coalescent in clusters (synangia). Homosporous. Mostly fleshy ferns of tropical forest. Extinct and living forms known; extinct tree ferns of this group in Lower Permian were large, important and dominant.
e.g. of fossil genera: *Asterotheca, Megaphyton, Psaronius.*
e.g. of living genera: *Angiopteris, Christensenia, Danaea, Marattia.*

Order Ophioglossales

Terrestrial, occasionally epiphytic. Leaf base more or less grasping, with thin stipules; leaves pinnately divided, simple, lobed or 2-ranked. Leaf divided into sterile and fertile segments and sporangia are borne on fertile segments only. Homosporous. Ferns small and fleshy. Living members only, with no early fossil record.
e.g. *Botrychium* (moonwort), *Helminthostachys, Ophioglossum* (adder's tongue).

C. Osmundidae

This group is intermediate between the eu- and leptosporangiate ferns, but does not necessarily link the two groups, since it has an almost continuous fossil history dating back to the Permian or Upper Carboniferous, and the survivors are 'living fossils'. The characteristics of the group are:
Sporangia not strictly leptosporangiate, although often described as such; not grouped into sori and have no annulus, or have only a primitive kind of annulus.

Order Osmundales

Terrestrial. Petiole with stipule-like dilations. Sporangia on axes with much-reduced segments or on unmodified blade. Homosporous. Living members usually large with massive stems.

e.g. of extinct genera: *Osmundites*, *Zalesskya*.

e.g. of living genera: *Osmunda* (royal fern), *Todea*.

D. Leptosporangiatae

(Filices, leptosporangiate ferns)

Sporangium leptosporangiate, i.e. sporogenous tissue derived from outer daughter cell. Sporangium wall, stalk and spores derived from outer daughter cell only, and adjacent cells not involved in their formation. Sporangium small; wall 1 cell thick, annulus usually present. Spore content low. Advanced.

Order Filicales

(Polypodiales)

Terrestrial, sometimes epiphytic. Leaves non-pulvinate and mostly non-stipulate, pinnately divided or simple. Sporangia on leaves, solitary or in sori. Homosporous. Mostly medium-sized. A very large group with extinct and living forms, and including most common ferns.

e.g. of extinct genera: *Alsophilites*, *Camptopteris*, *Coniopteris*, *Dictyophyllum*, *Gleichenites*, *Senftenbergia*.

e.g. of living genera: *Adiantum* (maidenhair fern), *Blechnum* (hard fern), *Dicksonia*, *Dryopteris* (male fern), *Gleichenia*, *Hymenophyllum* (filmy fern), *Matonia*, *Phyllitis* (hart's tongue), *Polypodium* (polypody), *Pteridium* (bracken), *Schizaea*.

Hydropterides

(Hydropterideae, water ferns)

This group of water ferns are leptosporangiate. They differ from the Filicales in the following ways:

Heterosporous. Micro- and megasporangia have no annulus, and are enclosed in special sporocarps at base of leaves.

There are two orders of water ferns:

Order Marsileales

Aquatic, with long rooted rhizomes. Leaves with long petiole, 2 to 4 terminal leaflets or none. Sporangia in hard, bean-like sporocarps, each sporocarp

probably representing a tightly folded pinna. Not known as fossils.
3 genera: *Marsilea* (water clover), *Pilularia* (pillwort), *Regnellidium*.

Order Salviniales

Aquatic, floating, not rooted to bottom. Leaves sessile. Sporangia in sac-like sporocarps, each sporocarp representing a single sorus whose indusium forms sporocarp wall. Not known as fossils.
2 genera: *Azolla* (mosquito fern), *Salvinia* (water spangles).

9　Spermatophyta: gymnosperms

Division Spermatophyta

(spermatophytes, seed plants)

Photosynthetic vascular plants which reproduce by seeds rather than by spores (Greek *sperma* = seed, *phyton* = plant), and show alternation of generations with very suppressed gametophyte generation. Sporophyte generation is dominant, conspicuous and independent. Gametophyte reduced, not the free-living generation, but spends its life on 1 or more sporophyte plants. Sporophyte consists of true roots, stems and leaves, all with vascular tissue; leaves with cuticle and stomata. Water not necessary for fertilization; male gamete moves into pollen tube which grows to female gamete by chemotropism (except in cycads and *Ginkgo* where sperm swims down pollen tube in chemotaxis). Heterosporous with microspores and megaspores; male gametophyte develops inside microspore (pollen grain) produced by meiosis inside microsporangium (pollen sac) attached to microsporophyll; female gametophyte develops inside megaspore (embryo sac) produced by meiosis inside a megasporangium (nucellus of ovule) attached to megasporophyll; archegonia absent in angiosperms but present in most gymnosperms in reduced form. Only microspores are dispersed, and are carried (in pollination) to the female structures which are retained on sporophyte. Vascular tissue confined to sporophyte and arranged in steles which are more advanced than those of pteridophytes. Secondary xylem present forming true wood. Great variety of leaf shape and venation. Sporangia arranged on sporophylls grouped together in cones in gymnosperms and into flowers in angiosperms.

Subgroups of spermatophytes

The Spermatophyta is often divided into two subdivisions, the gymnosperms and angiosperms (flowering plants) on the basis that in gymnosperms the seeds are borne on a megasporophyll which does not enclose them (Greek *gymnos* = naked), while in angiosperms the megasporophyll (carpel) encloses the seed (Greek *anggeion* = vessel) and forms a fruit.

It is now thought that the differences between the conifers and cycads are so great that the three groups should be given equal weight; so they may be considered as three equal subdivisions known as Coniferophytina (or Pinicae), Cycadophytina (or Cycadicae) and Magnoliophytina (or Angiospermae). Alternatively, it may be considered that the three groups, although very distinct, should still allow for the differences between angiosperms and gymnosperms. So the angiosperms and gymnosperms may be raised to division status as Pinophyta (gymnosperms) and Anthophyta or Magnoliophyta (angiosperms). Here, the terms angiosperms and gymnosperms are used.

Gymnosperms

(subdivision Gymnospermae or division Pinophyta)

Woody plants, shrubs to tall trees. Ovule naked (not enclosed in ovary). Female gametophyte with 500 or more cells or nuclei*. Female gametophyte produces definite but reduced archegonia*. Wood without vessels, only tracheids, and known as softwoods*. Phloem without companion cells*. Double fertilization absent. Endosperm made from female gametophyte and is haploid. One integument to ovule*, but integument divided into 3 layers.
* Except in some Gnetopsida.

Subgroups of gymnosperms

The gynmosperms may be divided into three subgroups. Where the gymnosperms are considered to be a subdivision, these groups are classes called Cycadopsida, Coniferopsida and Gnetopsida. Where the gymnosperms are considered to be a division these classes are raised to subdivision status as Cycadicae (Cycadophytina), Pinicae (Coniferophytina) and Gneticae (occasionally Gnetophytina). Where the differences between the cycads and conifers are thought very important and the two groups are given status equivalent to angiosperms, the names of the groups are Cycadophytina, Coniferophytina and Magnoliophytina; and the Gnetopsida are considered to be a class of Cycadophytina. Where cycads and conifers are raised to subdivision status, orders within each group are sometimes raised to class status.

The alternative systems of classification are summarized in Table 9.1.

For the purposes of this classification, the simpler system of three classes, Cycadopsida, Gnetopsida and Coniferopsida, is used (the right-hand column of the table), since differences between the groups can be described quite adequately at this level, and little purpose is served in confusing an already difficult situation by the introduction of further levels of classification. But the raising of the groups to higher levels is certainly important and relevant, and probably reflects evolutionary trends better than the simpler system.

The Cycadopsida and Coniferopsida appear at the same time in the fossil

Table 9.1: Summary of alternative systems of classification of gymnosperms

Subdivision Cycadophytina (Cycadicae)	Class Cycadopsida (Cycadatae)
Class Lyginopteridatae (Pteridospermae)	
Order Lyginopteridales (Cycadofilicales, Pteridospermales)	Order Lyginopteridales (Cycadofilicales, Pteridospermales) ⎤ Order Lygino-Pteridales
Order Caytoniales	Order Caytoniales ⎦
Class Bennettitatae	
Subclass Bennettitidae	
Order Bennettitales	Order Bennettitales
Subclass Pentoxylidae	
Order Pentoxylales	Order Pentoxylales
Class Cycadatae	
Order Cycadales	Order Cycadales ⎤ Order
Order Nilssoniales	Order Nilssoniales ⎦ Cycadales
Class Gnetatae or Subdivision Gneticae (Chlamydospermae)	Class Gnetopsida (Gnetatae, Chlamydospermae)
Subclass Welwitschiidae	
Order Welwitschiales	Order Welwitschiales ⎤
Subclass Ephedridae	
Order Ephedrales	Order Ephedrales Order Gnetales
Subclass Gnetidae	
Order Gnetales	Order Gnetales ⎦
Subdivision Coniferophytina (Pinicae, Coniferae)	Class Coniferopsida
Class Ginkgoatae	
Order Ginkgoales	Order Ginkgoales
Class Pinatae	
Subclass Cordaitidae	
Order Cordaitales	Order Cordaitales
Subclass Pinidae (Coniferae)	
Order Pinales	Order Pinales ⎤ Order
Order Voltziales	Order Voltziales ⎦ Coniferales
Subclass Taxidae	
Order Taxales	Order Taxales

record and may be diphyletic, i.e. have different origins (although opinion about this is changing as more fossil evidence appears). This would mean that the seed had arisen separately in each group. The Gnetopsida have almost no fossil record, and are possibly widely separated from the other groups.

Class Cycadopsida

Palm-like and fern-like plants. Leaves mostly compound and frond-like; basically pinnate in form and venation. Living species have motile sperm. Wood without vessels; wood manoxylic, except in *Pentoxylon*. Strobili, where present, simple; flower-like structures with perianth absent, except in Bennettitales. Ovules with 1 integument. Fertilization by pollen tube with 1 male nucleus. Seeds radially symmetrical. Extensive fossil record.

Order Lyginopteridales

(Cycadofilicales, Pteridospermales, seed ferns)

Extinct Palaeozoic and Mesozoic plants found from Devonian to Jurassic periods. Leaves usually fern-like, relatively large, pinnate, often several times. Ovules borne separately along margins of, or on surfaces of, pinnately compound megasporophylls. Megasporophylls not in strobili; like foliage leaves, or specialized structures, not leaf-like, sometimes peltate. Microsporophylls pinnately compound and not in strobili. Stems relatively slender, protostelic or polystelic.

e.g. *Glossopteris, Lyginopteris, Tetrastichia.*

Order Caytoniales

This order is sometimes included as a family (Caytoniaceae) in the Lyginopteridales. It differs from them in the following ways: Extinct Mesozoic plants found from the Triassic to Cretaceous periods. Leaves usually of 4 terminal leaflets arranged in 2 pairs, with short stalk between them, appearing palmate. Ovules borne in clusters, inside little cups on megasporophylls.

e.g. *Caytonanthus, Caytonia, Sagenopteris.*

Order Bennettitales

(Cycadeoideales)

Extinct Mesozoic plants from Triassic to Cretaceous. Leaves pinnately compound, or occasionally simple, with open (rarely closed) venation, and syndetocheilic stomata. Ovules stalked and borne on dome-shaped or elongated receptacle with scales which are united at end to form shield through which micropyle protrudes. No megasporophylls, but ovule-containing structure (gynoecium) sometimes subtended by perianth-like structure forming a 'flower', although perianth is sometimes called a bract. Microsporophylls sometimes pinnately compound, sometimes simple or reduced, sometimes surrounding megasporophylls forming hermaphrodite 'flowers'. Stems stout with wide pith, or slender.

e.g. *Bennettites (Cycadeoidea), Wielandiella, Williamsonia, Williamsoniella.*

Order Pentoxylales

Extinct Mesozoic plants found in Jurassic period, and probably shrubs or very small trees. Leaves thick, simple, lanceolate with diploxylic leaf trace and stomata formerly thought syndetocheilic, now thought haplocheilic; open venation; leaves borne on short shoots which also bear reproductive organs. Ovules sessile. Female organ like stalked mulberry, consisting of about 20 sessile seeds attached to central receptacle and surrounded by stony layer, then fleshy outer layer of integument uniting them. Microsporophylls form whorl of branched microsporangiophores fused basally into disc. Stems polystelic.

e.g. *Carnoconites, Nipaniophyllum, Nipanioxylon, Pentoxylon, Sahnia.*

This order is sometimes included as a family (Pentoxylaceae) in the Bennettitales, or its position is regarded as uncertain, but it shares some characteristics with both the Bennettitales and Cycadales, and has some unique features, e.g. the wood of *Pentoxylon* is pycnoxylic and resembles *Araucaria*.

Order Cycadales

(cycads)

Extinct and living plants found from Triassic, and most abundant in Mesozoic; woody, palm-like or fern-like. Leaves pinnately or (rarely) bipinnately compound with haplocheilic stomata and diploxylic leaf trace. Ovules borne on usually simple megasporophylls, each megasporophyll with a sterile tip and 2 to 8 ovules. Megasporophylls in strobili, except in *Cycas*, with cones terminal or lateral. Microsporophylls simple, scale-like or peltate, in strobili, with pollen sacs on lower side; sperm motile in living members, with spiral band of flagella. Stems usually unbranched, some with additional coaxial vascular cylinders; mucilage canals in pith and cortex. Seed larger than in other orders, and all living members are dioecious.

e.g. of extinct genera: *Palaeocycas.*

e.g. of living genera: *Cycas, Stangeria, Zamia.*

Order Nilssoniales

This order is often included as a family (Nilssoniaceae) in the Cycadales. It differs in the following ways:

All members extinct and Mesozoic only. Leaf trace not diploxylic. Female cone very lax and probably pendulous.

e.g. *Androstrobus, Beania, Nilssonia.*

Class Gnetopsida

(Gnetatae, Chlamydospermae)

Not like palms or ferns, but woody; trees, shrubs, lianas or stumpy turnip-like plants with stem below ground. Leaves simple, opposite or whorled, elliptic, strap-shaped or reduced to minute scales. Living species without motile sperm. Wood with vessels; anomolous, neither manoxylic nor pycnoxylic. Strobili compound, flower-like structures with perianth; whole strobilus often called an inflorescence, usually dioecious. Single erect ovule with 2 or 3 envelopes around it (often interpreted as integuments or as integuments + perianth) and a long micropyle forming a tube. Fertilization by pollen tube with 2 male nuclei. Seed variable. Almost no fossil record; probably widely removed from other 2 classes.

(The name Chlamydospermae means enclosed seed, from Greek *chlamys* = cloak, *sperma* = seed.)

In some classifications, all three members of the Gnetopsida are placed in the single order Gnetales, which has the characteristics of the class. In other systems, each genus is placed in its own order (sometimes raised to subclass status) with characteristics as below.

Order Welwitschiales

Plants with large, turnip-like, mostly underground stem. 2 thick, long, leathery, opposite leaves with parallel veins, persisting throughout life of plant, which split into ribbons as it ages; stomata syndetocheilic. Strobili cone-like. Archegonia absent. *Welwitschia*-like pollen is found from Permian, otherwise no fossils.

1 species: *Welwitschia mirabilis* (desert octopus).

Order Ephedrales

Usually branched shrubs with jointed stems, but a few are lianas. Leaves minute and scale-like. Strobili somewhat cone-like. Archegonia present. *Ephedra*-like pollen found from Permian, and *Ephedra* pollen from Eocene, but otherwise no fossils.

1 genus: *Ephedra*.

Order Gnetales

Woody plants, usually lianas, sometimes shrubs or trees. Leaves opposite with reticulate venation, like angiosperms in appearance; stomata formerly thought to be syndetocheilic, now thought haplocheilic. Strobili not cone-like. Archegonia absent. No fossil record. A few have companion cells in phloem, but these are different from those in angiosperms (see Glossary), and are not found in other Gnetopsida.

1 genus: *Gnetum*.

Class Coniferopsida

Not like palms or ferns, often tall trees. Leaves simple, basically dichotomous in form and venation, and often needle-shaped, paddle-like, scale-like or fan-like. Living species without motile sperm, except *Ginkgo*. Wood without vessels; wood pycnoxylic. Strobili simple or compound; no flower-like structures with perianth. Ovules with 1 integument. Fertilization by pollen tube with 1 male nucleus. Seeds bilaterally symmetrical. Extensive fossil record.

■■■ Order Ginkgoales

Trees known from Permian to present. Leaves strap-shaped or fan-shaped with dichotomous venation and leathery. No megastrobili, but 2 to 10 ovules borne at tip of branches or on almost unbranched axes; ovules form large seeds with flesh outer and stony inner layer. Microstrobili catkin-like, axillary, unbranched, bearing microsporangiophores with 2 to 12 pendulous microsporangia. Sperm motile with spiral band of flagella. Branching trees with long and short shoots except in earliest fossil members.
e.g. of extinct genera: *Baiera, Ginkgoites, Trichopitys*.
1 living species: *Ginkgo biloba* (maidenhair tree), a 'living fossil'.

■■■ Order Cordaitales

Extinct trees from Devonian to Permian periods. Leaves strap-shaped, grass-like or paddle-like, with parallel ventation, spirally arranged and up to 1 metre long. Megastrobili with sterile appendages below and ovule-bearing appendages above with 1 to 4 ovules. Microstrobili with sterile appendages below and fertile appendages above with 4 to 6 pollen sacs. Sperm unknown, but presence of pollen chambers suggests that motile sperm may have been formed. Tall trees with slender trunks and a crown of branches.
e.g. *Cordaites, Endoxylon, Poroxylon*.

■■■ Order Coniferales

(Pinales)

Trees or shrubs from Carboniferous to present. Leaves needle-like, scale-like or occasionally broad; spirally arranged or opposite, rarely whorled. Megastrobili often cone-like, with a main axis with numerous to few bract scales, each subtending or fused with 1 ovuliferous scale, bearing 2 to numerous ovules. Microstrobili usually cone-like with many scale-like microsporophylls and 2 to numerous fused or free pollen sacs. Sperm non-motile. Branching often with long and short shoots; resin canals in leaves, cortex and sometimes wood.
e.g. of extinct genera: *Lebachia, Voltziopsis, Walchia*.
e.g. of living genera: *Abies* (fir), *Araucaria* (Norfolk Island pine), *Cedrus* (cedar), *Cupressus* (cypress), *Juniperus* (juniper), *Pinus* (pine), *Podocarpus, Sequoia* (redwood), etc., many genera of living conifers.

The term conifer may be used just for this order or for the whole class. In some classifications, the two fossil families Lebachiaceae and Voltziaceae are separated into their own order, Voltziales:

Order Voltziales

This is a small order of fossil conifers, sometimes important in forming forests of the Upper Carboniferous, Permian, Triassic and Jurassic. They differ from the Coniferales in that the megastrobili were borne on short shoots.
e.g. *Lebachia, Voltziopsis.*

Order Taxales

Trees or shrubs from Triassic to present. Leaves needle-like, linear, spirally arranged. No megastrobili, but ovules solitary, surrounded by an aril, and terminal on short shoots. Microstrobili in small cones, scale-like or peltate, with 2 to 8 pollen sacs. Sperm non-motile. Densely branching evergreen shrubs or tall trees; no resin canals in wood or leaves.
1 extinct genus: *Palaeotaxus.*
e.g. of living genera: *Nothotaxus, Taxus* (yew), *Torreya.*

Progymnospermopsida

(Progymnospermae, progymnosperms)
A number of Devonian and Carboniferous fossils have been found which were thought to be pteridophytes (Primofilices), but are now thought to be nearer to gymnosperms although not of full gymnosperms status. These were grouped as a class of gymnosperms, the Progymnospermopsida, when connections were found between the fern-like fronds of *Archaeopteris* and the gymnosperm-like trunk of *Callixylon.* They are thought to form a link from the Psilophytopsida to the gymnosperms, and to represent the origins of both the Cycadopsida and Coniferopsida; but since they are not truely gymnosperms, they may still be classified as pteridophytes. They are placed in three orders:

Order Pityales

(equivalent to the pteridophyte Archaeopteridales)
Possibly ancestor of Coniferophytina. Moderately complex lateral organs. Xylem compact.
e.g. *Archaeopitys, Archaeopteris, Callixylon, Pitys.*

Order Aneurophytales

(equivalent to the pteridophyte Protopteridales)

Possibly ancestor of Cycadophytina. Complex pinnate lateral organs. Xylem dissected by small medullary rays and considered pycnoxylic.

e.g. *Aneurophyton, Palaeopitys, Tetraxylopteris.*

Order Protopityales

Possibly ancestral group of the other 2 orders; *Protopitys* is a very isolated genus, its wood suggesting affinities with Coniferopsida and its leaf traces with Cycadopsida; may be a pteridophyte exhibiting early stages of heterospory.

1 genus: *Protopitys.*

10 Spermatophyta: angiosperms

(Angiospermae, flowering plants, subdivision Magnoliophytina or division Magnoliophyta)

Trees, shrubs, or herbaceous plants; wide range of structure. Ovule enclosed in an ovary made of 1 or more carpels (megasporophylls) which after fertilization develop into fruit. Female gametophyte very reduced, with less than 500 cells, usually of about 8 nuclei only. Gametophyte does not produce archegonia. Wood usually contains vessels and tracheids, and known as hardwoods. Phloem contains companion cells, except in *Austrobaileya*. Double fertilization present. Endosperm made from triple fusion nucleus and is triploid. Usually 2 integuments to ovule.

The classification of angiosperms has undergone considerable revision recently, and familiar groups have been renamed to accord with the code of botanical nomenclature. Formerly the systems of Engler and Bentham and Hooker were widely used in textbooks and for plant arrangements in herbaria. Today the systems of the Armenian botanist Armen Takhtajan and the American Arthur Cronquist (which are similar and related to each other) are becoming accepted, but the arrangements are being modified all the time as more information comes to light. The system described here is based on Takhtajan and Cronquist.

Since the systems of Engler and Bentham and Hooker are still well known, the tables below relate the modern classification to them. The first table shows the modern arrangement and indicates some former groupings; beside each term there are codes referring to the Engler (E) and Bentham and Hooker (B&H) systems. The second table is an outline of these two systems with numbers and letters relating to the codes. In the modern system some of the orders have the same name, and comprise roughly the same families, as the earlier classifications, but other modern orders did not exist and were included in larger orders in former systems. Thus the code beside the modern Ranunculales in the Engler column is A18, meaning that the modern order Ranunculales was in Engler's system included in the 18th order of Archichlamydeae, which is the Ranales; the order Piperales is A2, meaning that the modern Piperales was included in Engler's second order of

Archichlamydeae, which is also called Piperales. If an order in the modern system has no number beside it, this means that, in former systems, its families were scattered amongst the other orders.

The modern system has seven subclasses of dicotyledons and four subclasses of monocotyledons, whose characteristics are described in the text. In the earlier classifications the monocots were not divided into subclasses but the dicots were, depending on whether the petals were fused or free. The group with free petals were called Archichlamydeae by Engler and Polypetalae by Bentham and Hooker. The plants with fused petals were called Sympetalae by Engler, Gamopetalae by Bentham and Hooker, and are known as Metachlamydeae in other systems. Bentham and Hooker had a third group, the Monochlamydeae or Incompletae, comprising those plants which had either no perianth or one whorl of usually sepaloid perianth. In Bentham and Hooker's system the dicots were thought ancestral to the monocots, and flowers with free petals more primitive than those with fused petals, the theory still held today. Engler thought that monocots were more primitive and placed them at the beginning of his system, but they are rearranged here for the sake of comparison.

The modern system considers that flowers with parts separate rather than fused, and with a spiral arrangement of parts rather than cyclic are primitive, and that fusion and reduction are more advanced.

Modern system after Cronquist and Takhtajan (former names and groupings shown in parentheses)

Class Magnoliopsida (Dicotyledons)

		E	B & H
I. Subclass Magnoliidae			
Order Magnoliales (Annonales)	⎤ Magnoliales	A18	P1
Order Laurales	⎦	A18	P1, MIV, MV
Order Piperales		A2	MIV
Order Aristolochiales	⎤ Aristolochiales	A15	MIII
Order Rafflesiales	⎦	A15	MIII
Order Nymphaeales		A18	P1
II. Subclass Ranunculidae		—	—
Order Illiciales		A18	P1
Order Nelumbonales	⎤ Ranales in part	A18	P1
Order Ranunculales	⎦	A18	P1
Order Papaverales (Rhoeadales in part)		A19	P2
Order Sarraceniales		A20	P2
III. Subclass Hamamelididae		—	—

Order Trochodendrales ⎤		A_{18}	P_1
Order Cercidiphyllales ⎬ Ranales in part		A_{18}	P_1
Order Eupteleales ⎦		A_{18}	P_1
Order Didymelales		—	—
Order Hamamelidales		A_{21}	P_{11}
Order Eucommiales		A_{21}	P_1
Order Urticales		A_{12}	MVII
Order Barbeyales		A_{12}	MVII
Order Casuarinales (Verticillatae)		A_1	MVII
Order Fagales ⎤ Fagales		A_{11}	MVII
Order Betulales ⎦		A_{11}	MVII
Order Balanopales		—	MVII
Order Myricales		A_5	MVII
Order Leitneriales		A_7	MVII
Order Juglandales		A_8	MVII
IV. Subclass Caryophyllidae		—	—
Order Caryophyllales (Chenopodiales, Centrospermae)		A_{17}	P_4
Order Polygonales		A_{16}	MI
Order Plumbaginales		S_3	G_5
Order Theligonales		A_{17}	MVII
V. Subclass Dilleniidae		—	—
Order Dilleniales ⎤		A_{27}, A_{21}	P_1
[Order Paeoniales]*		A_{18}	P_1
Order Theales ⎬ Parietales		A_{27}	P_5
Order Violales		A_{27}	P_2
Order Passiflorales ⎦		A_{27}	P_{13}
Order Cucurbitales		S_9	P_{13}
Order Begoniales		A_{27}	P_{13}
Order Capparales (Rhoeadales in part)		A_{19}	P_2
Order Tamaricales		A_{27}	P_4
Order Salicales		A_3	MVII
Order Ericales (Bicornes) ⎤ Ericales		S_1	G_4
Order Diapensiales ⎦		S_1	G_4
Order Ebenales		S_4	G6
Order Primulales		S_2	G_5

*Not formerly placed in the Parietales.

Order Malvales (Columniferae)	A26	P6
Order Euphorbiales (Tricoccae)	A23	MVII
Order Thymelaeales	A29	MV
VI. Subclass Rosidae	—	—
Order Saxifragales	A21	P11
Order Rosales	A21	P11
Order Fabales (Leguminosae)	A21	P11
Order Connarales	A21	P11
Order Nepenthales	A20	P11, MIII
Order Podostemales	A21	MII
Order Myrtales (Myrtiflorae)	A29	P12
Order Hippuridales (Haloragales)	A29	P11
Order Rutales ⎤ Terebinthales	A23	P7
Order Sapindales ⎦	A24	P10
Order Geraniales (Gruinales)	A23	P7
Order Polygalales	A23	P3
Order Cornales (Umbellales)	A30	P15
Order Celastrales	A24	P9
Order Rhamnales	A25	P9
Order Oleales (Contortae in part)	S5	G7
Order Santalales	A14	MVI
Order Elaeagnales	A29	MV
Order Proteales	A13	MV
VII. Subclass Asteridae	—	—
Order Dipsacales (Rubiales in part)	S8	G2
Order Gentianales (Rubiales in part & Contortae in part)	S8, S5	G7
Order Polemoniales (Tubiflorae in part)	S6	G8
Order Scrophulariales (Personatae) (Tubiflorae in part)	S6	G9
Order Lamiales (Tubiflorae in part)	S6	G10
Order Campanulales (Campanulatae)	S10	G3
Order Calycerales	S10	G2
Order Asterales	S10	G2
Class Liliopsida (Monocotyledons)	—	—
I. Subclass Alismidae (Helobiae)	2	—
Order Alismales (Alismatales)	2	VI

Order Hydrocharitales	2	I
Order Najadales	2	VI
II. Subclass Liliidae	—	—
Order Triuridales	3	VI
Order Liliales (Liliiflorae)	9	II, III
Order Iridales	9	II
Order Zingiberales (Scitamineae)	10	II
Order Orchidales (Microspermae, Gynandrae)	11	I
III. Subclass Commelinidae	—	—
Order Juncales	9	IV
Order Cyperales	4	VII
Order Bromeliales	8	II
Order Commelinales ⎤ Farinosae	8	III
Order Eriocaulales ⎦	8	VII
Order Restionales	8	VII
Order Poales (Glumiflorae)	4	VII
IV. Subclass Arecidae (Spadiciflorae)	—	—
Order Arecales (Principes)	5	IV
Order Cyclanthales (Synanthae)	6	V
Order Arales (Spathiflorae)	7	V
Order Pandanales	1	V
Order Typhales	1	V

Former systems of classification from Engler and Bentham and Hooker

Engler (E)   *Bentham and Hooker* (B&H)

Dicotyledoneae   Dicotyledones

  Archichlamydeae (A)   I. Polypetalae (P)

                                        Series I Thalamiflorae

    Order 1. Verticillatae

    Order 2. Piperales   Order 1. Ranales

    Order 3. Salicales   Order 2. Parietales

    Order 4. Garryales   Order 3. Polygalinae

    Order 5. Myricales   Order 4. Caryophyllinae

    Order 6. Balanopsidales   Order 5. Guttiferales

Order 7. Leitneriales

Order 8. Juglandales

Order 9. Batidales

Order 10. Julianiales

Order 11. Fagales

Order 12. Urticales

Order 13. Proteales

Order 14. Santalales

Order 15. Aristolochiales

Order 16. Polygonales

Order 17. Centrospermae

Order 18. Ranales

Order 19. Rhoeadales

Order 20. Sarraceniales

Order 21. Rosales

Order 22. Pandales

Order 23. Geraniales

Order 24. Sapindales

Order 25. Rhamnales

Order 26. Malvales

Order 27. Parietales

Order 28. Opuntiales

Order 29. Myrtiflorae

Order 30. Umbelliflorae

Order 6. Malvales

Series II Disciflorae

Order 7. Geraniales

Order 8. Olacales

Order 9. Celastrales

Order 10. Sapindales

Series III Calyciflorae

Order 11. Rosales

Order 12. Myrtales

Order 13. Passiflorales

Order 14. Ficoidales

Order 15. Umbellales

Sympetalae (S)

Order 1. Ericales

Order 2. Primulales

Order 3. Plumbaginales

Order 4. Ebenales

Order 5. Contortae

Order 6. Tubiflorae

Order 7. Plantaginales

Order 8. Rubiales

Order 9. Cucurbitales

II. Gamopetalae (G)

Series I Inferae

Order 1. Rubiales

Order 2. Asterales

Order 3. Campanales

Series II Heteromerae

Order 4. Ericales

Order 5. Primulales

Order 6. Ebenales

Series III Bicarpellatae

Order 10. Campanulatae

Order 7. Gentianales

Order 8. Polemoniales

Order 9. Personales

Order 10. Lamiales

III. Monochlamydeae or Incompletae (M)

Series I Curvembryae

Series II Multiovulatae Aquaticae

Series III Multiovulatae Terrestres

Series IV Micrembryae

Series V Daphnales

Series VI Achlamydosporeae

Series VII Unisexuales

Series VIII Anomalous families

Monocotyledoneae

Monocotyledons

Series I Microspermae

Order 1. Pandanales

Series II Epigynae

Order 2. Helobiae

Series III Coronarieae

Order 3. Triuridales

Series IV Calycinae

Order 4. Glumiflorae

Series V Nudiflorae

Order 5. Principes

Series VI Apocarpae

Order 6. Synanthae

Series VII Glumaceae

Order 7. Spathiflorae

Order 8. Farinosae

Order 9. Liliiflorae

Order 10. Scitamineae

Order 11. Microspermae

The system below is based on Cronquist and Takhtajan, but names from former systems are added where appropriate. Examples of families rather than genera are given here, and a common name or anglicized generic name of a member of the family is included where possible.

Class Magnoliopsida

(Magnoliatae, dicotyledons, dicots)

Embryo usually with 2 cotyledons. Leaves typically with reticulate venation, sometimes compound, seldom sheathed at base but stipules frequently present; petiole usually well developed. Bracteoles, where present, usually 2. In stem, vascular bundles usually borne in ring, with cambium persisting. Radicle generally persistent and becomes tap root; root cap and piliferous layer have common origin, except in Nymphaeales. Woody and herbaceous forms, but tree type thought to be more primitive. Floral parts (except carpels) usually in sets of 5 or 4 (seldom 3 or fewer). Nectaries of various types, often derived from stamens, rarely septal. Pollen grains usually tricolpate or tricolpate-derived.

Subclass Magnoliidae

Mostly woody. Some without vessels. Many with oil cells. Stomata usually with 2 subsidiary cells. Flowers mostly hermaphrodite; often spiral or spirocyclic. Gynoecium usually apocarpous. Pollen grains 2- or 3-celled, monocolpate or monocolpate-derived. Ovule with 2 integuments. Seed with much or little endosperm or none.

Order Magnoliales

(Annonales)

Trees or shrubs with alternate, simple leaves, and ethereal (aromatic) oils in all organs; Winteraceae have wood without vessels. Flowers spiral to cyclic, usually with free parts, hypogynous, usually hermaphrodite; perianth obvious but often not divided into petals and sepals. Stamens various in number. Ovary various, carpels numerous to few. Fruit various. Seed with abundant endosperm.

Comprises 8 families including Annonaceae (custard apple), Degeneriaceae (degeneria), Magnoliaceae (magnolia), Myristicaceae (nutmeg), Winteraceae (wintera).

This is thought to be the most primitive order of dicots. The next order, Laurales, is sometimes included in the Magnoliales.

Order Laurales

Usually trees, shrubs or climbers, with ethereal oils in cells; in *Austrobaileya*, phloem without companion cells, and in some Chloranthaceae xylem without vessels. Flowers hermaphrodite, unisexual or polygamous, hypogynous or epigynous, parts variable in number, in spirals or whorls. Stamens few to numerous. Ovary variable from apocarpous to syncarpous, with few to numerous ovules, often 1 per carpel. Fruit various. Seed with or without endosperm.

Comprises about 11 families including Austrobaileyaceae (austrobaileya), Calycanthaceae (calycanthus), Chloranthaceae, Lauraceae (laurel), Monimiaceae (monimia).

Order Piperales

Woody and herbaceous with simple leaves. Flowers hypogynous to epigynous, hermaphrodite or (rarely) unisexual, arranged in spikes and without perianth or without petals. Stamens 1 to 10. Ovary usually of 1 to 4 carpels, free or united, usually with 1 ovule (Piperaceae) or 2 to 10 ovules (Saururaceae). Fruit a drupe (often called a berry), follicle or capsule. Seed with dense mealy perisperm around endosperm and minute embryo. Comprises 2 families: Piperaceae (pepper), Saururaceae (lizard's tail).

The Chloranthaceae may be placed here rather than in the Laurales.

Order Aristolochiales

Herbs or climbing shrubs, often woody, sometimes with anomalous secondary thickening and with simple exstipulate leaves. Flowers hermaphrodite, usually zygomorphic and epigynous, with perianth parts in 3s, petals usually lacking and sepals often petaloid. Stamens 6 to numerous, free or united with style. Ovary syncarpous, made of sometimes 4 but usually 6 carpels, with numerous ovules and either 1 compartment and parietal placentation or 4 to 6 compartments and axile placentation. Fruit a capsule. Seed with fleshy endosperm.
Comprises 1 family: Aristolochiaceae (birthwort).

Order Rafflesiales

Body reduced to thalloid parasite. Flowers hermaphrodite or unisexual, epigynous, actinomorphic, with 1 whorl of petaloid perianth. Stamens 3 to numerous. Ovary syncarpous, of 3 to 20 carpels, containing many ovules. Fruit a berry. Seed with endosperm and perisperm.
Comprises 2 families: Hydnoraceae, Rafflesiaceae (rafflesia).

This order has been placed in the Santalales.

Order Nymphaeales

Herbaceous aquatic plants, with leaves not containing oil cells. Flowers hermaphrodite or unisexual, hypogynous or epigynous, very varied, especially in perianth. Stamens usually numerous. Ovary of 3 to numerous, free or united carpels. Fruit a follicle, nutlet or leathery berry. Seed with or without endosperm and perisperm.
Comprises 4 families including Ceratophyllaceae (hornwort) and Nymphaeaceae (water lily).

This order goes back to the Cretaceous period, and shows many features of ancestral monocots, which are thought to be derived from a common stock.

In some systems of classification, this order also includes the families Nelumbonaceae and Illiciaceae. Modern work suggests that these two families should be placed separately, the Nelumbonaceae in the order Nelumbonales, and the Illiciaceae, which shows many intermediate features, should be placed with the family Schisandraceae in the order Illiciales. This order is intermediate in character between the Magnoliidae and Ranunculidae, but here is placed as the first order of the subclass Ranunculidae.

Subclass Ranunculidae

Mostly herbaceous plants. Vessels present. Usually without oil cells. Stomata of various types, usually without subsidiary cells. Flowers hermaphrodite or unisexual; often spiral or spirocyclic. Gynoecium variable. Pollen grains usually 2-celled; tricolpate or tricolpate-derived. Ovule mostly with 2 integuments. Seed usually with much endosperm.

This subclass is near to the Magnoliidae but is more advanced. It is sometimes considered to be a superorder, Ranunculanae, within the Magnoliidae.

Order Illicidales

(Schisandrales)

Shrubs, sometimes climbing, or small trees, with simple exstipulate leaves. Flowers hermaphrodite or unisexual, hypogynous, actinomorphic, solitary, with 7 to numerous perianth segments. Stamens 4 to numerous. Ovary made of 5 to numerous free carpels with 1 to numerous ovules in each carpel. Fruit a group of follicles or drupe-like carpels. Seed with copious endosperm. Comprises 2 families: Illiciaceae (star anise), Schisandraceae (schisandra). See also Nymphaeales, above.

Order Nelumbonales

Large aquatic herbs with unique leaf arrangement, rhizome bearing 'triads' of 2 scale leaves and 1 foliage leaf. Flowers with numerous petals. Stamens numerous. Ovary of many carpels embedded separately in swollen spongy receptacle, each carpel containing 1 pendulous ovule. Fruit a collection of achenes in above receptacle. Seed without endosperm.
Comprises 1 family: Nelumbonaceae (Indian lotus).

This family is sometimes placed in the Nymphaeales and next to the Nymphaeaceae, but it differs from them in many important features and is best placed in its own order here.

Order Ranunculales

Herbaceous or woody plants, climbers or shrubs, usually with exstipulate leaves; alkaloids usually present. Flowers usually spiral, hermaphrodite, actinomorphic, but zygomorphic in some Ranunculaceae; perianth in 1 to 3 series, parts in 3s, 4s or 5s; petals secrete nectar. Stamens numerous. Ovary of several to many usually free carpels, rarely 1; ovules with marginal or sub-basal placentation. Fruit a berry or group of follicles or achenes. Seed usually with endosperm.

Comprises about 10 families including Berberidaceae (barberry), Lardizabalaceae (akebia), Menispermaceae (moonseed), Podophyllaceae (American mandrake), Ranunculaceae (buttercup).

The Paeoniales (now in the Dilleniidae) were formerly included here, but have several differences, e.g. do not have nectar-secreting petals (see Paeoniales).

Order Papaverales

Usually fleshy herbs, some shrubs or small trees, many containing milky or watery sap; with exstipulate leaves. Flowers cyclic (except sometimes stamens), hypogynous, hermaphrodite, actinomorphic or zygomorphic, usually with 2 whorls of perianth which falls off early; 2 to 4 sepals, 4 to 12 petals. Stamens 4, 2, or numerous. Ovary paracarpous with 2 to numerous fused carpels and usually numerous ovules. Fruit a capsule or made up of 1-seeded indehiscent joints. Seed with small embryo and fleshy, often oily endosperm and sometimes with aril.

Comprises 3 families: Fumariaceae (fumitory), Hypecoaceae, Papaveraceae (poppy).

Formerly the Papaverales and Capparales comprised the order Rhoeadales.

Order Sarraceniales

Carnivorous herbaceous plants growing from rhizome, usually with basal rosette of tubular or pitcher-like leaves modified for trapping animals. Flowers hermaphrodite, hypogynous, actinomorphic, with 1 to 2 whorls of perianth; sepals showy, usually 5, petals free, usually 5, or absent, yellow to purple. Stamens numerous. Ovary of 3 to 5 fused carpels; ovules numerous. Fruit a capsule. Seed minute, with endosperm.

Comprises 1 family: Sarraceniaceae (pitcher plants).

This order formerly included the Nepenthaceae and Droseraceae, and was placed in the Rosidae and known as the Nepenthales or Sarraceniales. But recent work on floral and pollen morphology suggests that the Sarraceniaceae show some very primitive features and should be placed near the Ranunculales. Similarities to the other two families lie mainly in that their leaves are modified as insect traps.

Subclass Hamamelididae

Mostly woody. Usually have vessels, except Trochodendrales. Oil cells absent. Stomata with 2 or more subsidiary cells, or without subsidiary cells. Flowers more or less reduced, mostly unisexual and often without petals; cyclic. Gynoecium usually coenocarpous. Pollen grains usually 2-celled; tricolpate or tricolpate-derived. Ovule mostly with 2 integuments. Seeds with much or little endosperm or none.

Many orders in this subclass are trees bearing catkins (aments) and were formerly placed in an artificial group called Amentiferae (catkin-bearing). This group included the Betulales, Fagales, Juglandales, Leitneriales, Myricales, and also the Salicales, which is rather different from the other orders and is now placed in the Dilleniidae.

Order Trochodendrales

Trees or shrubs with simple exstipulate leaves; vessels absent. Flowers without petals, but with calyx of 4 sepals in Tetracentraceae. Stamens 4 to numerous. Ovary of 4 to numerous carpels, free or partially fused, with several to numerous ovules. Fruit a capsule or group of follicles. Seed with oily endosperm.
Comprises 2 families: Tetracentraceae, Trochodendraceae.

This order is thought to be intermediate between the Magnoliales and Hamamelidales, but is nearer the latter and so placed in the Hamamelididae.

Order Cercidiphyllales

Trees with opposite leaves and united stipules. Flowers dioecious, with 4 sepals but no petals. Stamens 15 to 20. Ovary of 4 to 6 carpels with numerous ovules. Fruit a group of follicles. Seed with endosperm.
Comprises 1 family: Cercidiphyllaceae with 1 genus *Cercidiphyllum*.

This order is near both the Trochodendrales and Hamamelidales and is sometimes included in the latter.

Order Eupteleales

Trees or shrubs with exstipulate leaves. Flowers hypogynous, without perianth. Stamens numerous. Ovary of 6 to 18 free carpels, each containing 1 to 3 ovules. Fruit a cluster of samaras. Seed with oily endosperm.
Comprises 1 family: Eupteleaceae with 1 genus *Euptelea*.

This order is somewhat isolated systematically and has formerly been included in the Trochodendraceae.

Order Didymelales

Trees with simple leaves. Flowers dioecious, hypogynous. Male flowers subtended by 0 to 2 scales, with 2 sessile stamens. Female flowers subtended by 0 to 4 scales, with ovary of 1 carpel and 1 ovule. Fruit a large, 1-seeded drupe. Seed with copious endosperm.

Comprises 1 family: Didymelaceae with 1 genus *Didymeles*.

This order is sometimes included in the Hamamelidales.

Order Hamamelidales

Woody plants with simple or palmately lobed, stipulate leaves. Inflorescence condensed; flowers mostly small, often unisexual with reduction in floral structure. Stamens 4 to 14. Carpels free, or fused in region of ovary but with free styles. Fruit a capsule or nutlet. Seed with endosperm.

Comprises 3 families: Hamamelidaceae (witch hazel), Myrothamnaceae, Platanaceae (plane); and sometimes Altingiaceae, although this may be regarded as a subfamily of Hamamelidaceae.

Order Eucommiales

Trees with exstipulate leaves and latex. Flowers unisexual, naked, hypogynous, actinomorphic. Stamens 6 to 10. Ovary made of 2 fused carpels with 1 anatropous ovule. Fruit a samara. Seed with endosperm.

Comprises 1 family: Eucommiaceae with 1 genus *Eucommia*.

Order Urticales

Herbs, shrubs, vines or trees with usually simple stipulate leaves. Flowers hermaphrodite or unisexual, hypogynous, usually small, greenish, with perianth of 4 to 5 segments, alike and often sepaloid; sometimes only 1 perianth whorl or naked; usually actinomorphic. Stamens same number as, and appear before, perianth. Ovary of 1 or 2 fused carpels, usually containing 1 ovule. Fruit a nutlet, drupe, achene, samara or multiple. Seed with or without endosperm.

Comprises 4 families: Cannabaceae or Cannabinaceae (hemp), Moraceae (mulberry), Ulmaceae (elm), Urticaceae (nettle).

Order Barbeyales

Trees with simple, exstipulate leaves. Flowers dioecious, actinomorphic, without petals, but calyx of 3 to 4 sepals. Stamens 6 to 9. Ovary of 1 carpel and contains 1 anatropous ovule. Fruit dry, indehiscent, surrounded by membranous calyx. Seed without endosperm.

Comprises 1 family: Barbeyaceae with 1 genus *Barbeya*, which was formerly included in the Ulmaceae, or the Barbeyaceae was placed in the Urticales.

Order Casuarinales

(Verticillatae)

Trees or shrubs, with branches resembling a horsetail in habit. Flowers unisexual; male flowers in catkin-like spikes and female flowers in dense spherical heads at end of twigs. Male flowers with 2 median bract-like perianth lobes and several stamens hanging out of flower. Female flowers naked in axil of bract with 2 bracteoles, and have 2 thread-like stigmas and ovary of 2 fused carpels, with 2 compartments, 1 sterile, other with usually 2 ovules. Fruit made of whole head which becomes woody. Seed winged, still enclosed in woody bracteoles; without endosperm.

Comprises 1 family: Casuarinaceae (she oak).

Order Fagales

Trees or shrubs with ·simple stipulate leaves. Flowers monoecious, with reduced or absent perianth. Male flowers in catkins with 2 to many stamens. Female flowers epigynous, in catkins or occasionally in cone-like heads, or few, and appear before leaves; ovary surrounded by involucre of bracts, syncarpous, made from 2 to 3 carpels with 2 to 6 compartments, and 1 to 2 ovules in each comparment. Fruit a nut, sometimes winged. Seed without endosperm.

Comprises 3 families: Betulaceae (birch), Corylaceae (hazel), Fagaceae (beech).

In some systems of classification the Betulaceae and Corylaceae are placed in a separate order, the Betulales (see below).

Order Betulales

The families Betulaceae and Corylaceae may be placed here. Differences between the Fagales and Betulales are shown in the table below:

Fagales	Betulales
Both sexes have perianth	Perianth in 1 sex only
Styles 3 or more	Styles 2
Fruit large and nut-like, partly or completely enclosed in hard cup or shell	Fruit small, or large and nut-like, but cup (if present) is papery or leaf-like

Order Balanopales

Trees or shrubs with exstipulate leaves. Flowers dioecious, without perianth. Male flowers in catkins with 2 to 12 stamens. Female flowers solitary, subtended by many bracts; ovary of 2 to 3 fused carpels, with 2 to 3 compartments containing 2 ovules per compartment. Fruit an acorn-like drupe with 1 to 2 1-seeded pyrenes and persistent styles. Seed with little endosperm.

Comprises 1 family: Balanopaceae, of uncertain affinities.

This order may be included in the Fagales.

Order Myricales

Aromatic trees and shrubs with simple exstipulate leaves. Flowers usually unisexual and arranged in dense spikes surrounded by bracts, with no perianth in either sex. Male flowers with 2 to 16 stamens, usually 4, free or becoming joined. Female flowers with superior ovary made of 2 fused carpels, but with 1 compartment with 1 ovule. Fruit a drupe with waxy skin. Seed without endosperm.

Comprises 1 family: Myricaceae (sweet gale).

Order Leitneriales

Shrubs with resin canals and exstipulate leaves. Dioecious with flowers in spikes. Male flowers without perianth, and with 3 to 12 stamens. Female flowers with perianth of small scaly united segments, and ovary of 1 carpel with long style, ovary containing 1 ovule. Fruit drupe-like. Seed with thin endosperm.

Comprises 1 family: Leitneriaceae with 1 genus *Leitneria*.

Order Juglandales

Trees, often aromatic, with usually pinnate, exstipulate leaves. Flowers actinomorphic, either unisexual and arranged in catkins, or hermaphrodite and arranged in long slender spikes; perianth small, sepaloid, or absent. Male flowers with 3 to numerous stamens. Female flowers with inferior ovary in Juglandaceae and superior in Rhoipteleaceae; ovary of 2 fused carpels, with 1 compartment containing 1 ovule. Fruit a nut or drupe. Seed without endosperm.

Comprises 2 families: Juglandaceae (walnut), Rhoipteleaceae.

Subclass Caryophyllidae

Usually herbaceous, or small shrubs, rarely trees. Vessels always present. Oil cells absent. Stomata with 2 or 3, occasionally 4, subsidiary cells, or subsidiary cells absent. Flowers mostly without petals, hermaphrodite or unisexual, cyclic. Gynoecium apocarpous or coenocarpous. Pollen grains mostly 3-celled, tricolpate or tricolpate-derived. Ovule mostly with 2 integuments. Seed often with perisperm.

Order Caryophyllales

(Chenopodiales, Centrospermae)

Herbs, or rarely soft-wooded shrubs or trees, with cambium produced at various depths. Flowers usually hermaphrodite, and colour often due to betalains, pigments unique to this group; actinomorphic, hypogynous to perigynous; perianth in 1 to 2 whorls, the outer sepaloid and inner petaloid or sepaloid. Stamens usually 3 to 10. Ovary usually syncarpous and of 1 to about 20 carpels, 1-celled (or 2- to several-celled) with placentation axile to free-central or basal. Fruit various. Seed usually with perisperm.

Comprises about 10 to 16 families including Aizoaceae (carpetweed), Amaranthaceae (amaranth), Basellaceae (Madeira vine), Cactaceae (cacti), Caryophyllaceae (pink), Chenopodiaceae (goosefoot), Didiereaceae (didierea), Nyctaginaceae (4 o'clock plant), Phytolaccaceae (pokeweed), Portulacaceae (purslane).

The Cactaceae may be separated into its own order, Cactales, or may be placed here because its members contain betalains. If separated, its characteristics are as follows:

Order Cactales

(Opuntiales)

Stems succulent, often bearing spines in areoles (clearly defined areas developed from apical bud), leaves usually absent or reduced. Flowers hermaphrodite, epigynous. Stamens numerous, spirally arranged. Carpels 3 to 20 in whorl, with 1 style; stigmas as many as carpels; ovary usually with 1 compartment, numerous ovules. Fruit a berry, or variable. Seed with perisperm.

Comprises 1 family: Cactaceae (cacti).

Order Polygonales

Herbs or shrubs, rarely trees; leaves often with sheathing stipules (ocrea). Flowers hermaphrodite, hypogynous, actinomorphic; perianth segments usually 6, sepaloid or petaloid, free or united. Stamens usually 6. Ovary of 3 carpels and 1 compartment with 1 ovule. Fruit a triangular nut. Seed with endosperm.

Comprises 1 family: Polygonaceae (buckwheat).

This order is sometimes included in the Caryophyllales.

Order Plumbaginales

Herbs or small shrubs, with exstipulate leaves. Flowers hermaphrodite, hypogynous, actinomorphic; calyx commonly 5-lobed, often strongly ribbed and membranous between lobes; petals 5, usually fused. Stamens 5, somewhat epipetalous. Ovary of 5 fused carpels and 1 compartment, with 5 styles and 1 ovule with basal placentation. Fruit a nut. Seed usually with mealy endosperm.

Comprises 1 family: Plumbaginaceae (leadwort).

Order Theligonales

Herbs with fleshy stipulate leaves. Flowers unisexual. Male flowers with perianth of 2 to 5 segments, stamens usually 7 to 12. Female flowers with 3 to 4 perianth segments and ovary of 1 carpel containing 1 ovule. Fruit a nut-like drupe. Seed with endosperm.

Comprises 1 family: Theligonaceae (or Thelygonaceae), also called Cynocrambaceae.

This family was formerly included in the Caryophyllaceae, or may be included in the Haloragales.

Subclass Dilleniidae

Woody and herbaceous. Vessels always present. Oil cells absent. Stomata of various types, usually without subsidiary cells. Flowers hermaphrodite or unisexual, usually with sepals and petals; in primitive families, perianth spiral or spirocyclic. Gynoecium apocarpous, or more frequently coenocarpous. Pollen grain usually 2-celled, tricolpate or tricolpate-derived. Ovule mostly with 2 integuments. Seed usually with endosperm.

Order Dilleniales

Woody plants (often climbing) with often leathery leaves. Flowers hermaphrodite, actinomorphic, with 5 to numerous sepals and usually 5 petals. Stamens usually numerous. Ovary of 1 to numerous carpels. Fruit various, often with aril. Seed with endosperm.

Comprises 2 families: Crossosomataceae, Dilleniaceae (dillenia).

Order Paeoniales

Mostly herbs, some shrubs, with large deeply-cut leaves. Flowers large, actinomorphic, hermaphrodite, with 5 sepals, 5 to 10 free petals and a fleshy disc. Stamens many, dehiscing to inside of flower. Carpels 2 to 5. Fruit 2 to 5 follicles, usually with many seeds. Seed large, at first red, later black, arillate, with copious endosperm.

Comprises 1 family: Paeoniaceae (peony).

This family was formerly included in the Ranunculales, or even within the family Ranunculaceae. It differs from the Ranunculaceae in the possession of a disc, inward dehiscence of the stamens, large arillate seeds, distinctive anatomy, and in that the petals do not secrete nectar. The Paeoniales is now sometimes included in the order Dilleniales.

Order Theales

Trees, shrubs or woody climbers, rarely herbs; leaves simple and evergreen. Flowers actinomorphic, hermaphrodite, with 4, 5 or many sepals, sometimes like bracts and bracteoles; petals usually 5, free or rarely fused. Stamens usually numerous, often united in bundles or in a ring. Carpels usually 5, united, or when free joined by united styles; ovules with axile placentation. Fruit various. Seed with or without endosperm.

Comprises 21 families including Clusiaceae or Guttiferae (mangosteen), Dipterocarpaceae, Hypericaceae (St. John's wort), Theaceae (tea).

Order Violales

(Cistales)

Shrubs or small trees, less often herbs, with stipulate leaves. Flowers usually hermaphrodite, hypogynous, actinomorphic to zygomorphic; sepals 3 to 15, petals usually 5. Stamens usually 5 to numerous. Ovary syncarpous with 1, or 3 to 5, compartments, and ovules usually with parietal placentation. Fruit usually a capsule or sometimes fleshy. Seed with endosperm.

Comprises 9 families including Cistaceae (rock rose), Flacourtiaceae, Violaceae (violet).

In some systems, the next order, Passiflorales, is included in the Violales. In other arrangements, the Cistaceae is not part of the Violales, but with the Hypericaceae make up the order Cistiflorae.

Order Passiflorales

Shrubs, herbs, climbers or small trees; leaves variable. Flowers hermaphrodite or unisexual, hypogynous, variable. Stamens various, free. Ovary of 3 to 5 fused carpels and several to many ovules usually with parietal placentation. Fruit a capsule or berry. Seed with endosperm.

Comprises 5 families including Caricaceae (pawpaw) and Passifloraceae (passion flower). In some systems it also includes the family Tamaricaceae, although this may be placed in its own order, see below.

This order may be included in the Violales.

Order Cucurbitales

Herbs or small trees, often climbing by tendrils, and frequently with bicollateral vascular bundles; leaves usually large and deeply lobed or compound. Flowers unisexual, epigynous, usually actinomorphic and showy; calyx lobed; petals free or fused. Stamens 1 to numerous, free or united, sometimes epipetalous. Ovary with 1 to 4 compartments, ovules usually numerous with parietal or axile placentation. Fruit a capsule or berry (pepo). Seed with little or no endosperm.

Comprises 1 family: Cucurbitaceae (gourds).

This order is sometimes included in the Violales.

Order Begoniales

(Datiscales)

Mostly succulent herbs, but also shrub-like herbs and large trees; leaves variable. Flowers usually unisexual; sepals 2 to 9, often petaloid; petals 2 to 8 or more or none. Stamens 4 to 25 or more. Carpels 3, rarely 5 or 2, styles 2 to 5, ovules numerous. Fruit a capsule, rarely a berry. Seed small with reticulate coat and little or no endosperm.

Comprises 2 families: Begoniaceae (begonia), Datiscaceae.

This order is sometimes included in the Violales.

Order Capparales

Herbs, shrubs, small trees and lianas; leaves variable, with stipules. Flowers hermaphrodite, hypogynous or perigynous, actinomorphic or zygomorphic. Stamens numerous to few. Ovary of 2 or more fused carpels, and ovules with parietal placentation. Fruit a fleshy berry or drupe, or dry dehiscent siliquas, siliculas or capsules. Seed with little endosperm.

Comprises 8 families including Brassicaceae or Cruciferae (mustard, etc.), Capparaceae or Capparidaceae (capers), Resedaceae (mignonette).

Order Tamaricales

Trees, shrubs or rarely herbs, with leaves small and scale-like or heather-like. Flowers usually small, hermaphrodite, hypogynous, actinomorphic; 4 to 7 sepals and 4 to 7 petals. Stamens 5 to 10 or numerous. Ovary syncarpous, of 3 to 5 carpels, with 1 compartment and numerous ovules with parietal or basal placentation. Fruit a capsule. Seed with or without endosperm.

Comprises 3 families: Fouquieriaceae (ocotilla), Frankeniaceae (sea heath), Tamaricaceae (tamarisk).

This order is sometimes included in the Violales or Passiflorales.

Order Salicales

Trees or shrubs with simple, usually stipulate leaves. Flowers dioecious, hypogynous, in catkins, often appearing before leaves and having very small or absent perianth. Stamens 2 to many. Ovary usually of 2 fused carpels, containing 1 compartment and numerous ovules with parietal placentation.

Fruit a capsule. Seed without endosperm.

Comprises 1 family: Salicaceae (willow, popular).

This order was formerly placed in the Amentiferae (see p. 123) but it is very different from the other catkin-bearing plants.

Order Ericales

(Bicornes)

Shrubs, rarely trees or herbs, some herbaceous saprophytes or root parasites, with simple exstipulate, often leathery leaves. Flowers hermaphrodite, rarely unisexual, hypogynous or epigynous, actinomorphic or zygomorphic; sepals and petals usually each 4 to 10, free or united. Stamens usually twice as many as sepals or petals. Ovary usually of 2 to 7 fused carpels, with 2 to 7 compartments and 1 to many ovules in each compartment, with axile placentation. Fruit a capsule, berry or drupe. Seed with abundant endosperm.

Comprises 10 families including Actinidiaceae (actinidia), Clethraceae (pepperbush), Empetraceae (crowberry), Ericaceae (heath), Monotropaceae (Indian pipe), Pyrolaceae (wintergreen).

Order Diapensiales

Mainly arctic and alpine evergreen undershrubs with rosettes of simple leaves. Flowers with 5 sepals and 5 petals. Stamens 5, often also 5 staminodes. Ovary of 3 fused carpels and numerous ovules with axile placentation. Fruit a capsule. Seed with endosperm.

Comprises 1 family: Diapensiaceae.

This order is often included in the Ericales.

Order Ebenales

Trees or shrubs with simple leaves. Flowers hypogynous, actinomorphic, sepals and petals each 4 or 5 and fused. Stamens variable in number, arranged in 1 or more whorls. Ovary of many carpels and ovules with axile placentation. Fruit a drupe or berry. Seed with endosperm.

Comprises 5 families including Ebenaceae (ebony), Sapotaceae (sapote), Styracaceae (storax).

Order Primulales

Herbs, shrubs or trees, with exstipulate leaves variously arranged. Flowers usually hermaphrodite, hypogynous or rarely perigynous, actinomorphic or rarely zygomorphic; parts usually in 5s, calyx and corolla both 4- to 9-lobed. Stamens as many as and opposite corolla lobes, joined to tube. Ovary of 5 fused carpels, with 1 style and 1 compartment containing 2 to many ovules with free-central placentation. Fruit a capsule, drupe or berry. Seed with endosperm.

Comprises 3 families: Myrsinaceae (myrsine), Primulaceae (primrose), Theophrastaceae.

Order Malvales

(Columniferae)

Trees, shrubs or herbs with stipulate leaves, often mucilaginous. Flowers hermaphrodite or sometimes unisexual, hypogynous, usually actinomorphic; calyx and corolla usually with parts in 5s, petals sometimes absent. Stamens numerous, free or united into 1 or more bundles by fusion of filaments, or united into column around style (hence Columniferae). Ovary syncarpous, of 2 or more carpels and compartments, and ovules with axile placentation. Fruit various. Seed usually with endosperm.

Comprises 9 families including Bombacaceae (baobab, silk cotton), Malvaceae (mallow, cotton), Sterculiaceae (cocoa), Tiliaceae (lime).

In some classifications the next two orders, the Euphorbiales and Thymelaeales, are included in the Malvales, and the order may be placed in the Rosidae rather than Dilleniidae.

Order Euphorbiales

(Tricoccae)

Trees, shrubs or occasionally herbs, with usually stipulate leaves, sometimes reduced. Flowers usually unisexual, hypogynous, actinomorphic, with sepals but without petals. Stamens 1 to numerous, free or united. Ovary of 2 or 3 compartments, ovules with axile placentation. Fruit a capsule or drupe. Seed usually with abundant endosperm.

Comprises about 7 families including Buxaceae (boxwood), Euphorbiaceae (spurge), Simmondsiaceae (jojoba).

There is some disagreement about the families in this order, suggesting a polyphyletic origin. Sometimes the order is not recognized, or comprises only the Euphorbiaceae, or may be placed in the Rosidae.

Order Thymelaeales

Mostly shrubs with exstipulate leaves. Flowers usually hermaphrodite, hypogynous, actinomorphic, with parts in 4s or 5s, and receptacle hollowed out forming a tube-like structure; sepals petaloid; corolla conspicuous or absent. Stamens as many, twice, or half number of sepals. Ovary of usually 1 to 2 carpels, with as many compartments as carpels, each with 1 ovule. Fruit usually an achene, berry or drupe, often enclosed in persistent receptacle. Seed with little or no endosperm.

Comprises 1 family: Thymelaeaceae (mezereon).

This order may be placed in the Rosidae, sometimes as a family of the order Myrtales.

Subclass Rosidae

Woody or herbaceous. Vessels always present. Oil cells absent. Stomata of various types, with 2 subsidiary cells or none. Flowers mostly hermaphrodite, with sepals and petals, or without petals. Gynoecium apocarpous or coenocarpous. Pollen grain usually 2-celled, tricolpate or tricolpate-derived. Ovule mostly with 2 integuments. Seed with or without endosperm.

Order Saxifragales

Trees, shrubs, lianas and herbs, with variable leaves. Flowers mainly hermaphrodite, sometimes unisexual, actinomorphic or rarely weakly zygomorphic, hypogynous, perigynous or epigynous, usually with calyx and corolla; petals free or joined. Stamens few to numerous. Carpels free or fused; ovule with 2 integuments, or more rarely 1. Fruit a drupe, capsule, follicle, nut or berry. Seed usually with abundant endosperm.

Comprises 28 families including Crassulaceae (orpine), Cunoniaceae, Escalloniaceae (escallonia), Grossulariaceae (gooseberry), Hydrangeaceae (hydrangea), Parnassiaceae (grass of Parnassus), Saxifragaceae (saxifrage).

This order is sometimes included in the Rosales.

Order Rosales

Usually woody plants with variable leaves; often with stipules. Flowers usually with parts in 5s, hypogynous to epigynous, often perigynous. Stamens numerous and free. Ovary of 1 to numerous carpels, free or united. Fruit various. Seed usually with endosperm.

Comprises 3 families: Chrysobalanaceae (coco plum), Neuradaceae, Rosaceae (rose).

Order Fabales

(Leguminosae)

Trees, shrubs, herbs or vines, often with pinnate or bipinnate leaves, sometimes trifoliate or simple, often with stipules. Flowers hermaphrodite, hypogynous to perigynous, actinomorphic to zygomorphic, with 5 sepals and usually 5 petals. Stamens often 10, sometimes numerous, often united into 1 or 2 bundles. Ovary of 1 carpel, containing several ovules with marginal or parietal placentation. Fruit usually a legume (pod), sometimes highly modified. Seed usually with little or no endosperm.

Comprises 3 families: Caesalpiniaceae (peacock flower), Mimosaceae (mimosa), Fabaceae or Papilionaceae (beans etc.).

In earlier systems, the Leguminosae was of family status, and the families here were regarded as subfamilies. Leguminosae is the ICBN permitted alternative name of the Fabaceae.

Order Connarales

This order is closely related to the Fabales. It differs from them in the following ways:

Leaves without stipules. Ovary of more than 2 carpels, each with 2 ovules. Fruit usually 1 follicle with 1 seed, with or without endosperm, but with an aril.

Comprises 1 family: Connaraceae.

This order is sometimes included in the Sapindales.

Order Nepenthales

Herbaceous carnivorous plants with leaves adapted for trapping animals. Flowers hermaphrodite or unisexual, hypogynous to perigynous, actinomorphic; sepals 4 to 5, united at base; petals 5, sometimes absent. Stamens 4 to numerous, free or partly fused. Ovary of 3 to 5 fused carpels, containing few to numerous ovules with axile or parietal placentation. Fruit usually a capsule. Seed with endosperm.

Comprises 2 families: Droseraceae (sundew), Nepenthaceae (pitcher plants).

This order formerly included the family Sarraceniaceae and was known as Sarraceniales. But it is now thought that the Sarraceniaceae has substantial differences from the Nepenthaceae, and the Sarraceniaceae is placed in a separate order, Sarraceniales, in the Ranunculidae. The Droseraceae is clearly advanced of the Sarraceniaceae, but the position of the Nepenthaceae is uncertain, although its pollen and ovule morphology places it with the Droseraceae.

Order Podostemales

(Podostemonales)

Plants living in rushing water and growing on rocks in rivers; vegetative organs very varied, often filamentous, ribbon-like or lichen-like. Flowers simple, hermaphrodite, hypogynous, naked, enclosed in spathe. Stamens 1 to numerous. Ovary usually of 2 fused carpels and 2 compartments, containing numerous ovules with axile placentation. Fruit a capsule. Seed without endosperm.

Comprises 1 family: Podostemaceae.

Order Myrtales

(Myrtiflorae)

Mostly shrubs or trees, usually with simple leaves, with internal phloem in some families. Flowers usually actinomorphic, hermaphrodite, parts in 4s or 5s, perigynous to epigynous. Stamens usually free. Ovary syncarpous with 1 to many ovules. Fruit usually a capsule, drupe or berry. Seed without endosperm.

Comprises 13 families including Combretaceae (myrobalan), Lythraceae (loosestrife), Myrtaceae (myrtle), Onagraceae (evening primrose), Punicaceae (pomegranate), Rhizophoraceae (mangrove), Trapaceae (water chestnut).

Order Hippuridales

(Haloragales)

Land, marsh or water plants of various habits. Flowers inconspicuous, hermaphrodite or unisexual, perianth 4 + 4, or 4 or 0. Stamens 4 + 4, or fewer, or 1. Ovary of 1 to 4 carpels with 1 to 4 compartments and 1 ovule in each compartment. Fruit an achene, nut, or drupe. Seed with endosperm.

Comprises 3 families: Gunneraceae (gunnera), Haloragaceae or Haloragidaceae, Hippuridaceae (mare's tail).

Order Rutales

Trees or shrubs, rarely herbs, with leaves often compound and frequently gland-dotted. Flowers mostly hermaphrodite, hypogynous or weakly perigynous, usually actinomorphic; sepals usually free; petals usually free. Stamens as many or twice as many as petals. Ovary syncarpous, made of 1 to 20 carpels, containing 1 to 5 compartments, with usually 1 or 2 ovules in each compartment. Fruit various. Seed with or without endosperm.

Comprises 12 families including Anacardiaceae (cashew), Burseraceae, Meliaceae (mahogany), Rutaceae (citrus, rue), Simaroubaceae.

Order Sapindales

(Acerales)

Trees, shrubs or lianas, rarely herbs, with usually pinnate, exstipulate leaves. Flowers polygamous or dioecious, hypogynous to slightly perigynous, actinomorphic or zygomorphic, usually small; sepals 3 to 5; petals 3 to 5, often clawed, rarely absent. Stamens often twice as many as petals. Ovary of 2 to 5 but usually 3 carpels, mainly syncarpous, containing 1 to 2 ovules in each compartment with axile placentation. Fruit various but basically a capsule. Seed usually without endosperm.

Comprises 9 families including Aceraceae (maple), Greyiaceae (wild bottle-brush), Hippocastanaceae (horse chestnut), Sapindaceae (soapberry).

This order and the Rutales were formerly grouped together as Terebinthales.

Order Geraniales

(Gruinales)

Trees, shrubs, or herbs, with usually stipulate leaves. Flowers usually hermaphrodite, hypogynous, actinomorphic to zygomorphic; sepals free or somewhat fused; petals often clawed, occasionally joined or absent. Stamens usually twice as many as petals. Ovary syncarpous, with 3 to 5 compartments and ovules 1 to few in each compartment and with axile placentation. Fruit various but rarely fleshy. Seed usually without endosperm.

Comprises 20 families including Balsaminaceae (balsam), Erythroxylaceae (coca), Geraniaceae (geranium), Linaceae (flax), Malpighiaceae, Oxalidaceae (wood sorrel), Tropaeolaceae (nasturtium), Zygophyllaceae (caltrop).

The Linaceae and Erythroxylaceae with the Humiriaceae may be separated into the order Linales.

Order Polygalales

Herbs, shrubs or small trees, whose leaves have stipules small or absent. Flowers hermaphrodite, hypogynous to subperigynous, usually zygomorphic; sepals 5, often unequal; petals 1 to 5, free or sometimes joined. Stamens up to 12, sometimes only 1 fertile, sometimes joined. Ovary syncarpous with 1 to 3 compartments and ovules with axile or apical placentation. Fruit a capsule, drupe or samara. Seed with or without endosperm.

Comprises 5 families: Krameriaceae, Polygalaceae (milkwort), Tremandraceae, Trigoniaceae, Vochysiaceae.

Order Cornales

(Apiales, Araliales, Umbellales)

Trees, shrubs, herbs, rarely lianas, with leaves often much divided. Flowers usually small, hermaphrodite, unisexual or polygamous, epigynous, actinomorphic to weakly zygomorphic, usually small and arranged in umbels, cymes or racemes, with parts usually in 4s or 5s, occasionally up to 10; sepals usually partly fused; petals free, rarely absent. Stamens usually the same number as petals. Ovary of fused carpels, with 1 to 2, sometimes many compartments, with 1 ovule in each compartment. Fruit a berry, drupe or schizocarp. Seed with much or little endosperm.

Comprises 14 families including Apiaceae or Umbelliferae (carrot), Araliaceae (ginseng), Cornaceae (dogwood), Davidiaceae (paper handkerchief tree), Garryaceae (garrya), Nyssaceae (nyssa).

In some systems this order is divided into two, the Cornales which includes the Cornaceae and Garryaceae, and the Araliales which includes the Araliaceae and Umbelliferae.

Order Celastrales

Trees, shrubs or vines, occasionally herbs, with simple leaves. Flowers usually hermaphrodite, rarely unisexual, hypogynous to perigynous, usually small; sepals usually 4 to 5; petals 4 to 5, free, rarely absent. Stamens often 4 to 5, alternating with petals. Ovary syncarpous, usually containing 1 to 2 ovules in each compartment, with axile or apical placentation. Fruit various. Seed usually with abundant endosperm.

Comprises about 12 families including Aquifoliaceae (holly), Celastraceae (staff tree), Hippocrateaceae (hippocratea), Icacinaceae (icacina).

Order Rhamnales

Trees, shrubs or woody climbers, rarely herbs, with usually stipulate leaves. Flowers similar to Celastrales but stamens alternate with sepals. Fruit usually a drupe or berry. Seed with endosperm.

Comprises 3 families: Leeaceae, Rhamnaceae (buckthorn), Vitaceae (grape).

Order Oleales

(Ligustrales)

Trees, shrubs or climbers with simple or pinnate exstipulate leaves. Flowers usually hermaphrodite, actinomorphic, hypogynous with 4 fused sepals and usually 4 fused petals. Stamens usually 2, epipetalous, alternating with petals. Ovary of 2 fused carpels, with 2 compartments containing usually 2 ovules. Fruit usually 1-seeded, a drupe, berry, capsule or samara. Seed with or without endosperm.

Comprises 1 family: Oleaceae (olive and privet).

Order Santalales

Usually woody, often parasitic on other angiosperms or rarely on gymnosperms; leaves simple, entire, sometimes scale-like, exstipulate. Flowers hermaphrodite or unisexual, epigynous, actinomorphic; calyx and corolla with parts in 4s or 5s, calyx lobed or reduced, petals present or absent, sometimes united into tube. Stamens same number as calyx lobes. Ovary with 1 compartment and few ovules with axile placentation. Fruit a drupe or berry, sometimes a nut or achene. Seed with endosperm.

Comprises 13 families including Balanophoraceae, Loranthaceae (loranthus), Santalaceae (sandalwood), Viscaceae, formerly included in the Loranthaceae (mistletoe).

Order Elaeagnales

Much-branched shrubs, often with leathery leaves, covered with scaly hairs and often with thorns. Flowers hermaphrodite or unisexual, without petals, hypogynous, with parts in 2s or 4s. Stamens as many or twice as many as sepals. Ovary of 1 carpel, containing 1 ovule. Fruit drupe-like. Seed with little or no endosperm.
Comprises 1 family: Elaeagnaceae (oleaster).
   This order is sometimes included in the Proteales.

Order Proteales

Xerophytic shrubs and trees with entire or much divided exstipulate leaves with thick cuticle and hairs to reduce transpiration. Flowers very showy with pollen freely exposed, hypogynous, usually hermaphrodite, often zygomorphic, with petaloid perianth of 4 fused segments, with segments rolled or bent back. Stamens 4, epipetalous, with only anthers free. Ovary of 1 carpel, containing many or few ovules (or one); style long, bent inwards. Fruit a follicle, capsule, drupe or nut. Seed without endosperm.
Comprises 1 family: Proteaceae (protea).

Subclass Asteridae

Usually herbaceous. Vessels always present. Oil cells absent. Stomata of various types, usually without subsidiary cells. Flowers usually hermaphrodite and cyclic. Gynoecium coenocarpous (paracarpous). Pollen grain 3- or 2-celled, tricolpate or tricolpate-derived. Ovules small with single integument. Seed with or without endosperm.

Order Dipsacales

Herbs or shrubs, seldom small trees; leaves exstipulate. Flowers actinomorphic or zygomorphic, epigynous, hermaphrodite or unisexual, occasionally sterile; calyx absent or poorly developed; epicalyx sometimes well developed in fruit; corolla of fused petals, 3- to 5-lobed. Stamens epipetalous, usually 3 to 5. Ovary syncarpous, with 2 to 5 compartments, each with 1 to many ovules. Fruit various. Seed with or without endosperm.
Comprises 4 families: Adoxaceae (moschatel), Caprifoliaceae (honeysuckle), Dipsacaceae (teasel), Valerianaceae (valerian).

Order Gentianales

Trees, shrubs or herbs, usually with simple opposite leaves; phloem usually to inside of xylem. Flowers hermaphrodite and usually hypogynous; corolla of 4 to 5 fused petals. Stamens usually 4 to 5. Ovary of 2 carpels, ovules with parietal or axile placentation; ovules usually numerous with 1 integument and a 1-layered nucellus. Fruit usually a berry, capsule or follicle. Seed often winged, with or without endosperm.

Comprises 11 families including Apocynaceae (dogbane), Asclepiadaceae (milkweed), Gentianaceae (gentian), Loganiaceae (logania), Menyanthaceae (bog bean), Rubiaceae (madder).

In modern systems the Rubiaceae may be removed to an order of its own, the Rubiales. Formerly this was an order which included the modern Dipsacales and most of the Gentianales.

Order Polemoniales

Typically trees, shrubs, vines or herbs, a few parasitic. Flowers usually hypogynous, with fused petals, actinomorphic, with parts in 5s. Stamens 5 or numerous. Ovary of 1 to 5, mostly 2 carpels, usually fused. Fruit a capsule or of 1 to 4 nutlets. Seed with or without endosperm.

Comprises 8 families including Boraginaceae (borage), Convolvulaceae (morning glory), Cuscutaceae (dodder), Hydrophyllaceae (water leaf), Polemoniaceae (phlox).

Order Scrophulariales

(Personatae)

Trees, shrubs, vines, herbs, sometimes parasitic; leaves various. Flowers actinomorphic to zygomorphic, hypogynous or rarely epigynous; with 4 to 6 sepals and petals, petals usually fused. Stamens 5, 4, or 2, epipetalous. Ovary of 2 or 5 fused carpels, usually with 1 to 2 compartments, ovules 1 to many with a single integument and axile or parietal placentation. Fruit usually a capsule or berry, rarely a schizocarp, nut or drupe. Seed with or without endosperm.

Comprises 18 families including Acanthaceae (acanthus), Bignoniaceae (jacaranda), Buddleiaceae (buddleia), Gesneriaceae (African violet), Lentibulariaceae (butterwort), Plantaginaceae (plantain), Scrophulariaceae (antirrhinum), Solanaceae (nightshade).

Order Lamiales

Herbs, shrubs or trees with usually simple leaves. Flowers hypogynous, usually hermaphrodite, zygomorphic, with parts in 5s; petals fused. Stamens 2, 4 or 5, epipetalous. Ovary usually of 2 carpels and 4 compartments with 1 ovule in each compartment and usually a forked style. Fruit of 4 nutlets, or a drupe that separates into 4 pyrenes, or rarely a capsule. Seed with or without endosperm.

Comprises 4 families: Callitrichaceae (water starwort), Lamiaceae or Labiatae (mint etc.), Phrymaceae (lopseed), Verbenaceae (verbena).

Order Campanulales

(Campanulatae)

Usually herbs, often with latex vessels and mostly simple exstipulate leaves. Flowers hermaphrodite, usually epigynous, actinomorphic to zygomorphic, calyx and corolla 5-lobed, corolla sometimes 2-lipped. Stamens as many as corolla lobes, anthers sometimes united or touching. Ovary usually of 2 to 5 compartments, with usually numerous ovules in each compartment, sometimes 1. Fruit various. Seed with endosperm.

Comprises 7 families including Campanulaceae (bellflower), Goodeniaceae, Lobeliaceae (lobelia).

The Compositae has been placed here, which may be correct as it has similar chemicals, and the touching of the anthers and inferior ovary suggest a close relationship.

Order Calycerales

Herbs with exstipulate leaves. Flowers in heads with involucre of bracts, hermaphrodite or unisexual, epigynous, parts in 4s to 6s. Stamens with united filaments. Ovary of 1 carpel with 1 ovule. Fruit an achene. Seed with little endosperm.

Comprises 1 family: Calyceraceae.

This order is sometimes placed in the Campanulales, but the unilocular ovary and 1 pendulous ovule make it sufficiently different to warrant an order of its own.

Order Asterales

Herbs, shrubs or rarely trees or woody climbers with exstipulate leaves. Flowers hermaphrodite or unisexual, epigynous, actinomorphic to zygomorphic, usually crowded in heads, surrounded by bracts; calyx small, often with thread-like lobes (pappus); corolla usually 4- to 5-lobed. Stamens epipetalous, with 5 or 4 anthers joined to each other. Ovary of 2 carpels but 1 compartment, containing 1 ovule with basal placentation. Fruit an achene. Seed without endosperm.

Comprises 1 family: Asteraceae or Compositae (daisy etc.).

Class Liliopsida

(Liliatae, monocotyledons, monocots)

Embryo with 1 cotyledon, or embryo undifferentiated. Leaves usually with parallel venation, often sheathed at base; petiole seldom developed. Bracteoles, when present, usually 1. In stem, vascular bundles generally scattered (or occasionally in 2 or more rings), usually without cambium. Radicle of short duration, and soon replaced by adventitious roots; root cap and piliferous layer of different origins. Herbaceous and tree-like (primarily herbaceous). Floral parts (except carpels) typically borne in sets of 3 (seldom 4, never 5). Nectaries chiefly occur in septa between carpels (i.e. of septal type). Pollen grains monocolpate or monocolpate-derived, never tricolpate.

Subclass Alismidae

(Alismatidae, Helobiae)

Herbaceous, aquatic or partly aquatic. Vessels absent or in roots only. Oil cells absent. Stomata when present with 2, occasionally 4, subsidiary cells. Flowers hermaphrodite or unisexual. Gynoecium usually apocarpous. Pollen grain mostly 3-celled, monocolpate or 2-polyporate. Ovule with 2 integuments. Seed with little or no endosperm.

This subclass is equivalent to the order Helobiae in Engler's system, the term helobial referring to endosperm development in which the endosperm disappears.

Order Alismales

(Alismatales)

Herbs living in water or wet places, sometimes marine. Flowers hypogynous, actinomorphic, hermaphrodite or unisexual; perianth in 2 whorls, usually 3 sepals and 3 petals. Stamens 3, 6 or numerous. Ovary apocarpous with 1 to numerous carpels and numerous ovules. Fruit an achene or follicle. Seed without endosperm.

Comprises 3 families: Alismaceae (water plantain), Butomaceae (flowering rush), Limnocharitaceae.

This order is related to the Nymphaeales amongst the dicots.

Order Hydrocharitales

Aquatic herbs in water or wet places, sometimes marine. Flowers actinomorphic, usually dioecious, epigynous, arranged in spathe of 1 or 2 bracts with male flowers numerous, female flowers solitary; perianth segments in 2 whorls, usually 3 to each series. Stamens 1 to 3 to numerous. Ovary of 2 to 15 fused carpels, with 1 compartment and numerous ovules with parietal placentation. Fruit various. Seed without endosperm.

Comprises 1 family: Hydrocharitaceae (frog's-bit).

Order Najadales

(Potamogetonales)

Herbs living in water or wet places, sometimes marine; leaves linear, with scales in axils. Flowers hypogynous, hermaphrodite or unisexual; perianth absent or of 1 whorl, sometimes 2 similar whorls. Stamens usually 1 to 6. Ovary of few (often only 1) free or joined carpels with usually 1 ovule with basal or apical placentation. Fruit usually dry. Seed with little or no endosperm.

Comprises 10 families including Najadaceae or Naiadaceae, Potamogetonaceae (pondweed), Scheuchzeriaceae, Zosteraceae (eel-grass).

Subclass Liliidae

Herbaceous or occasionally tree-like. Vessels mostly in roots, rare in stems and leaves. Oil cells absent. Stomata usually without subsidiary cells, occasionally with 2 (some Liliales) or 2 or more (Zingiberales). Flowers hermaphrodite or unisexual. Gynoecium usually coenocarpous (mostly syncarpous), occasionally almost apocarpous. Pollen grain usually 2-celled, most monocolpate. Ovule with 2 integuments. Seed with or without endosperm.

This subclass is sometimes extended to include the next subclass, the Commelinidae.

Order Triuridales

Saprophytes with scale leaves. Small, long-stalked hermaphrodite or unisexual hypogynous flowers with 1 petaloid perianth whorl of 3 to 8 segments. Stamens 3, 4, or 6. Female flowers with 2 staminodes, and ovary apocarpous, of many carpels, each with basal ovule. Fruit a collection of achenes with thick wall. Seed with endosperm.

Comprises 1 family: Triuridaceae.

This order is sometimes placed in the Alismidae.

Order Liliales

(Liliiflorae)

Mostly herbaceous plants with rhizomes or bulbs, some shrubs or trees with unusual secondary thickening, some xeromorphic, succulent or climbers. Flowers usually hermaphrodite, hypogynous to epigynous, actinomorphic or zygomorphic, parts in 3s or multiples of 3; perianth 3 + 3, free or united, petaloid or sepaloid. Stamens 3 + 3 or fewer. Ovary of 3 fused carpels, usually of 3 compartments and ovules with axile placentation. Fruit usually a capsule or berry. Seed with endosperm.

Comprises about 20 families including Agavaceae (agave), Alliaceae (onion), Alstroemeriaceae, Amaryllidaceae (daffodil), Dioscoreaceae (yam), Liliaceae (lily), Pontederiaceae (water hyacinth), Smilacaceae (smilax).

In some classifications, the next order, Iridales, is included in the Liliales.

Order Iridales

Annual or perennial herbs; leaves basal or on definite stem. Flowers hermaphrodite, actinomorphic or zygomorphic, epigynous; perianth 3 + 3, petaloid. Stamens 3 to 6, fused to perianth. Ovary of 3 carpels, and of 3 compartments and ovules with axile placentation, or of 1 compartment and ovules with parietal placentation; ovules few to many. Fruit a capsule. Seed with much or little endosperm.

Comprises 4 families: Burmanniaceae, Corsiaceae, Geosiridaceae, Iridaceae (iris).

Order Zingiberales

(Scitamineae)

Tropical herbs with rhizomes, sometimes large and woody. Flowers zygomorphic, epigynous, with 1 or 2 perianth whorls arranged in 3s. Stamens usually 3 + 3, but often reduced, even to 1 stamen, with sterile staminodes. Ovary usually with 3 compartments, ovule large, and 1 to many in each compartment. Fruit usually a berry or capsule, sometimes highly modified. Seed with endosperm, often with aril.

Comprises 8 families including Cannaceae (canna), Marantaceae (arrowroot), Musaceae (banana), Strelitziaceae (bird of paradise flower), Zingiberaceae (ginger).

This order may be placed in the Commelinidae.

Order Orchidales

Perennial herbs, often with tubers or with stems swollen into pseudobulbs, often epiphytes or occasionally saprophytes; leaves simple, frequently rather thick. Flowers mostly hermaphrodite, epigynous, zygomorphic; perianth of 2 whorls, usually both petaloid, sometimes outer sepaloid. Stamens 2 or 1, fused to style to form column; pollen usually agglutinated into masses (pollinia). Ovary of 3 carpels and usually 1 compartment, often twisted through 180°, with numerous ovules and parietal placentation. Fruit usually a capsule. Seed minute, without endosperm and with an undifferentiated embryo.

Comprises 1 family: Orchidaceae (orchids).

Subclass Commelinidae

Mostly herbaceous. Vessels in all organs or restricted to roots. Oil cells absent. Stomata with 2, 4 or 6 subsidiary cells, occasionally without subsidiary cells. Flowers mostly hermaphrodite with somewhat reduced perianth. Gynoecium syncarpous or paracarpous. Pollen grain 2- or 3-celled, monocolpate or monoporate. Ovule mostly with 2 integuments. Seed usually with mealy endosperm.

This subclass is sometimes included in the Liliidae. Many of the orders here were included in the order Farinosae in Engler's system, from the Latin *farina* = flour, referring to the mealy endosperm present in many groups.

Order Juncales

Herbs with long narrow channelled or grass-like leaves. Flowers in heads, hypogynous, actinomorphic, hermaphrodite, perianth 3 + 3, green or brown. Stamens free, 3 + 3, or inner whorl missing. Ovary syncarpous, of 3 fused carpels, with 1 or 3 compartments and ovules 3 to many. Fruit a capsule. Seed with mealy endosperm.

Comprises 2 families: Juncaceae (rush), Thurniaceae.

Order Cyperales

Mostly perennial herbs with rhizomes and usually solid stems triangular in section; leaves usually linear and sheathing at base, sometimes reduced to sheaths. Flowers hermaphrodite or unisexual, hypogynous, small, crowded on heads or spikes each subtended by bract; perianth of scales, bristles or absent. Stamens usually 3. Ovary of 3 or 2 carpels and 1 compartment with 1 ovule. Fruit dry, indehiscent. Seed with endosperm.

Comprises 1 family: Cyperaceae (sedge).

In some systems, the Gramineae is included in the Cyperales.

Order Bromeliales

Terrestrial and epiphytic with reduced stem and rosette of fleshy water-absorbing and water-storing leaves, forming a funnel full of water, with inflorescences in centre surrounded by coloured bracts. Flowers hypogynous to epigynous, hermaphrodite, sometimes functionally unisexual, actinomorphic, parts in 3s; perianth of 3 sepaloid and 3 petaloid segments. Stamens 6, often epipetalous. Ovary of 3 fused carpels, with 3 compartments, 1 style and 3 stigmas, and a large number of ovules with axile placentation. Fruit a berry or capsule, if capsule, seeds light and winged. Seed with mealy endosperm.

Comprises 1 family: Bromeliaceae (pineapple).

In some systems, the Bromeliales is included in the Liliidae, next to the Zingiberales.

Order Commelinales

Herbs, variable in vegetative form. Flowers hypogynous, hermaphrodite, perianth in 2 whorls of 3, outer sepaloid and inner petaloid. Stamens 3, 6 or absent. Ovary of 3 fused carpels, with few to numerous ovules with parietal or free basal placentation. Fruit a capsule. Seed with endosperm, often mealy. Comprises 4 families including Commelinaceae (tradescantia), Rapateaceae, Xyridaceae (xyris).

Order Eriocaulales

Herbs with narrow leaves. Flowers small, unisexual, arranged in heads, hypogynous; perianth membranous or scale-like in 2 whorls, inner often united. Stamens 2 to 6. Ovary of 3 or 2 compartments with 1 ovule in each compartment. Fruit a capsule. Seed with mealy endosperm.
Comprises 1 family: Eriocaulaceae (pipe-wort).

Order Restionales

Xeromorphic, tufted or climbing plants, leaves usually sheathing. Flowers dioecious, hypogynous, sometimes monoecious or hermaphrodite; perianth 3 + 3, sometimes absent. Stamens 6, 3 or 2. Ovary of 1 to 3 carpels, containing 1 to 3 compartments with 1 ovule in each. Fruit a capsule, nut or drupe-like. Seed with mealy endosperm.
Comprises 4 families including Restionaceae (restio), Flagellariaceae.

Order Poales

(Glumiflorae)

Annual or more often perennial herbs, rarely woody, stems often hollow at internodes; leaves usually linear, with ligule, and sheathing at base. Flowers (florets) in inflorescences of various types but always with distinctive subunit, the spikelet, bearing empty bracts; florets hermaphrodite or unisexual, hypogynous, small, usually enclosed in 2 bracts; perianth represented by 2 minute scales (lodicules). Stamens often 3. Ovary of 1 to 3 carpels, with 1 compartment and 1 ovule often joined to side of carpel, and 2 feathery stigmas. Fruit usually a caryopsis. Seed with endosperm.
Comprises 1 family: Poaceae or Gramineae (grasses).
    This order is sometimes included in the Cyperales.

Subclass Arecidae

(Spadiciflorae)

Herbaceous or tree-like plants, most with broad leaves with petioles. Vessels in all organs or restricted to roots (Arales). Oil cells absent. Stomata with 2, 4 or occasionally 6 subsidiary cells. Flowers mostly unisexual, small and numerous, grouped into inflorescences generally subtended by spathe-like bracts; perianth of 6 segments, reduced to bristles, scales or absent.

Gynoecium syncarpous or paracarpous, but occasionally apocarpous in some palms. Pollen grain 2-celled, rarely 3-celled, mostly monocolpate. Ovule large with 2 integuments. Seed usually with endosperm.

Order Arecales

(Principes)

Tree-like or sometimes climbing plants, leaves feather- or fan-like. Flowers in compound spikes or spike-like racemes, usually in spathe; actinomorphic and unisexual, calyx and corolla usually 3. Stamens usually 6. Ovary usually of 3 carpels, syncarpous with 3 compartments or apocarpous. Fruit a berry or drupe. Seed with endosperm.

Comprises 1 family: Arecaceae or Palmae (palms).

Order Cyclanthales

(Synanthae)

Often palm-like plants, climbers or large herbs. Monoecious with male and female flowers alternating over surface of spadix. Male flowers naked or with thick short perianth and 6 to numerous stamens. Female flowers naked or with 4 fleshy perianth segments and long thread-like staminode in front of each. Ovary of 2 to 4 fused carpels; ovaries sunk in spadix and united; numerous ovules. Multiple fruit with numerous seeds. Seed with endosperm.

Comprises 1 family: Cyclanthaceae.

Order Arales

(Spathiflorae)

Herbs, occasionally woody climbers or floating aquatics. Flowers very small, hermaphrodite or unisexual, hypogynous, densely crowded on spadix, or rarely few together; inflorescence sometimes more or less enclosed in large bract (spathe); perianth small or absent. Stamens 1 to 6. Ovary with 1 to many compartments, ovule placentation various. Fruit usually a berry. Seed with or without endosperm.

Comprises 2 families: Araceae (arum), Lemnaceae (duckweed).

Order Pandanales

Mostly sea coast or marsh plants with tall stems supported by aerial roots, and some climbers, with stem twisted so leaves seem to run in spirals (hence screw pines); leaves usually with thorny margin. Inflorescence a racemose spadix with unisexual flowers. Male flowers with numerous stamens. Female flowers with numerous carpels, apocarpous or syncarpous, with 1 or many compartments, or ovary reduced to 1 carpel or a row of carpels, with sessile stigma (no style); ovules anatropous. Fruit a berry or multilocular drupe. Seed with oily endosperm.

Comprises 1 family: Pandanaceae (screw pines).

Order Typhales

Marsh or aquatic herbs with rhizomes and linear leaves, sheathing at base. Flowers unisexual, hypogynous, small, densely crowded on spikes or heads; perianth small, sepaloid, often of scales or threads. Stamens 2 or more. Ovary of 1 carpel and compartment with 1 ovule. Fruit dry. Seed with endosperm. Comprises 2 families: Sparganiaceae (bur reed), Typhaceae (cat–tail).

This order is sometimes placed in the Commelinidae.

Appendix: life cycles with alternation of generations

(showing the rise to dominance and independence of the sporophyte generation)

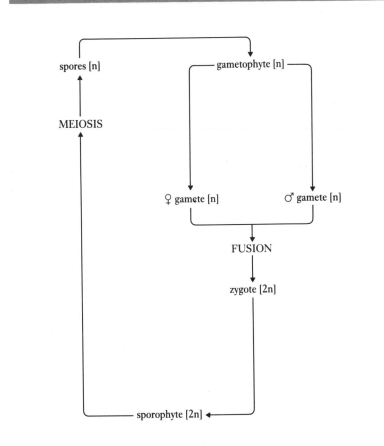

generalized life cycle

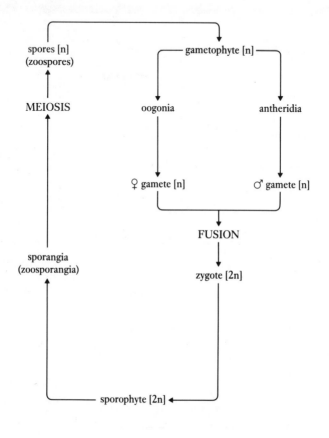

generalized life cycle in algae

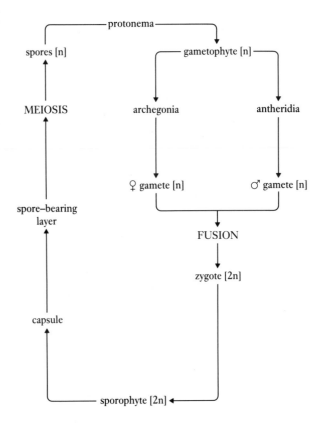

generalized life cycle in bryophytes

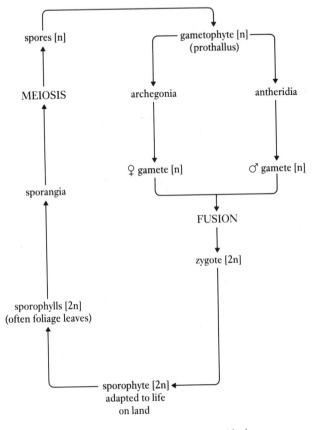

generalized life cycle in homosporous pteridophytes

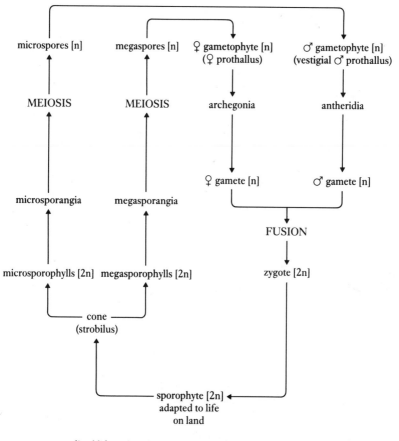

generalized life cycle in heterosporous pteridophytes

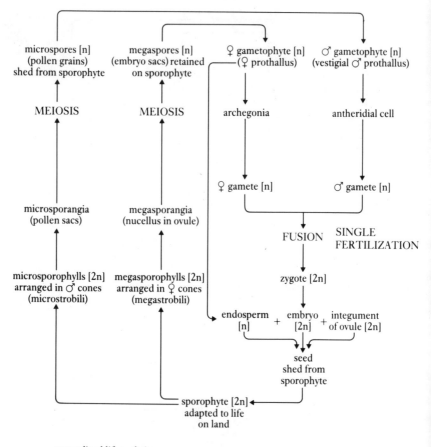

generalized life cycle in gymnosperms

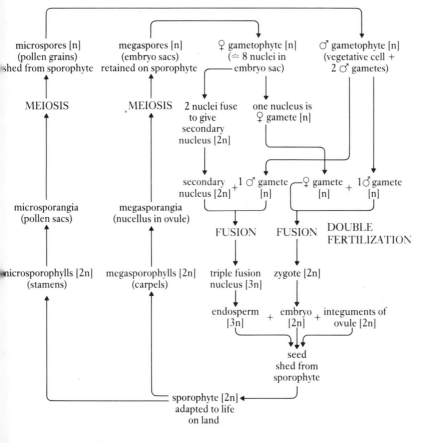

generalized life cycle in angiosperms

Glossary

acervulus: a dense, cushion-like mass of conidiophores and conidia formed by some fungi; *plu.* acervuli.

achene: a small, dry indehiscent, single-seeded fruit.

acrocarpous: bearing archegonia, and hence the sporophyte generation, at the tip, and usually of erect growth.

actinomorphic: radially symmetrical.

adventitious root: a root arising from a place other than the radicle of a seed.

aecidiospore: a spore formed in an aecidium.

aecidium: a cup-shaped body containing chains of aecidiospores; *plu.* aecidia.

akinete: a type of non-motile spore.

alternation of generations: the condition of having 2 types of plant body in a life cycle, a haploid gametophyte generation producing gametes by mitosis and a diploid sporophyte generation producing spores by meiosis, each generation giving rise to the other, so that the generations alternate in the life cycle.

amphigastrium: one of the underleaves in a liverwort; *plu.* amphigastria.

anatropous: of or having ovule bent over against the stalk.

androecium: collectively, the stamens of a flower.

anisogamy: the fusion of gametes differing only in size and not in general form; *a.* anisogamous; see also physiological anisogamy.

annulus: special thick-walled cells making up part of the opening mechanism of a fern sporangium, usually in the form of a ring.

anther: the part of the stamen containing the pollen grains.

antheridium: a male sex organ, containing male gametes; *plu.* antheridia.

aperturate: having pores.

apical: at the tip; *n.* apex; *plu.* apices.

aplanosporangium: a sporangium in which aplanospores develop.

aplanospore: a non-motile spore.

apocarpous: of or having an ovary with carpels free from one another.

apophysis: the lower part of the capsule in mosses.

apothecium: an open, cup-shaped fruit body (or a club-shaped derivative of it) containing ascospores.

arborescent: tree-like in growth form.

archegonium: a female sex organ, typically flask-shaped and consisting of a neck, made of neck cells and containing neck canal cells, and a swollen venter containing the ovum; *plu.* archegonia.

aril: an outgrowth of a seed coat; *a.* arillate.

arthrospore: a spore resulting from the fragmentation of a hypha.

ascocarp: a fruit body containing asci, in ascomycetes.

ascogenous: ascus-producing.

ascospore: one of the spores produced by meiosis inside an ascus.

ascostroma: a structure on which asci are borne; *plu.* ascostromata.

ascus: a sporangium containing ascospores; *plu.* asci.

asporogenous: without spores.

autospore: a daughter cell formed within an algal cell and having all the characteristics of its parent in miniature before being set free.

autotrophic: able to build up organic food material from inorganic substances by photosynthesis or chemosynthesis.

auxiliary cell: a specialized vegetative cell in some red algae, into which the zygote nucleus migrates.

auxospore: a resting spore in diatoms, formed from the zygote after sexual reproduction, and which restores the normal size of the organism.

axile chloroplast: one lying in the axis of the containing cell.

axile placentation: position of ovules in the ovary, where they lie on placentae in the angles formed by the meeting of septa in the middle of the ovary.

basidiocarp: a fruit body bearing basidia, in basidiomycetes.

basidiospore: one of the (usually) 4 spores formed by meiosis on the outside of a cell called a basidium.

basidium: the cell (or occasionally row of cells) which bears basidiospores; *plu.* basidia.

berry: a fleshy fruit, usually several-seeded, without a stony layer surrounding the seeds.

bicollateral: having internal as well as external phloem in vascular bundle.

bilateral symmetry: having only one plane of symmetry, i.e. only forming mirror images if cut in half one way; in flowers also called zygomorphy.

biloproteins: characteristic pigments in algae which contain a protein part; formerly called phycobilins, sometimes biliproteins.

bipinnate: of or having a compound leaf with its leaflets also pinnately divided.

bitunicate: having a wall of 2 layers.

bract: a leaf-like structure subtending an inflorescence, a flower, a group of sporangia, or a group of sex organs.

bracteole: a very small leaf-like structure on a flower stalk.

budding: the production of daughter cells as rounded outgrowths of the parent cell.

calcified: impregnated with calcium salts.

calyptra: the cap on a moss capsule, being the remains of the archegonium.

calyx: the sepals collectively.

cambium: a layer of cells, not at the tip, in a stem or root, capable of dividing to form secondary tissues such as wood.

capillitium: a tangled mass of thread-like structures mixed with spores.

capsule: in bryophytes, the part of the sporophyte containing the spores; in angiosperms, a dry dehiscent fruit composed of more than one carpel.

carbonaceous: blackened and hard, as though charred.

carotenes: a group of hydrocarbons producing a yellow colour.

carotenoids: a group of yellow, orange or red plant pigments including the carotenes and xanthophylls.

carpel: one of the units of which the gynoecium of angiosperms is composed, being equivalent to a megasporophyll of pteridophytes and gymnosperms.

carpogonium: a female sex organ in red algae; *plu.* carpogonia; *a.* carpogonial.

carpospore: a spore formed after fertilization of an ovum in a carpogonium.

carposporophyte: a parasitic generation that bears carpospores.

cartilaginous: resembling cartilage in consistency.

caryopsis: the fruit in grasses, being 1-seeded, with the fruit wall and seed coat united.

catkin: an inflorescence of reduced sessile flowers arranged on a common stalk, usually either all male or all female.

cellulin: a substance, somewhat like cellulose, in hyphae of some aquatic fungi.

chemolithotrophic, chemosynthetic: using carbon dioxide as the source of carbon and obtaining energy by oxidation of inorganic compounds.

chemotaxis: the swimming of a free-living part of a plant, e.g. spermatozoid, in relation to a chemical stimulus.

chemotropism: the growth of a plant structure, e.g. pollen grain, in relation to a chemical stimulus.

chlamydospore: a thick-walled spore.

chrysolaminarin: a polysaccharide storage produce in some algae: also called leucosin.

cilium: in bryophytes, a thread-like structure between peristome teeth; *plu.* cilia.

clamp connection: a hypha across the septum in some basidiomycetes.

cleistothecium: a completely enclosed ascocarp without a pore for release of ascospores.

coccoid: spherical; *n.* coccus; *plu.* cocci.

coenobium: a motile colony of cells arranged in a constant and reproducible way; *plu.* coenobia.

coenocarpous: with fused carpels.

coenocytic: filamentous but without cross walls (septa); also called aseptate.

colonial: living as a group of cells organized into a cluster, usually with little or no division of labour.

columella: central internal stalk of sporophore; central sterile tissue in bryophyte capsule.

commensal: an organism living with another and not harming the host.

companion cell: in angiosperms only, a nucleated cell associated with a sieve tube element and derived from the same parent cell; similar cells occur in some Gnetales, but do not originate from the same parent cell as the sieve tube element.

cone: an arrangement of microsporophylls and/or megasporophylls on the same axis, forming a more-or-less conical structure; also called strobilus.

conidium: a non-sexual spore; *plu.* conidia.

conidiophore: a hypha bearing conidia.

conjugation: the fusion of morphologically similar gametangia.

corolla: the petals collectively.

cortex: the region below the epidermis.

cotyledon: one of the first leaves of the plant, present in the seed as part of the embryo, and either remaining within the seed coat or emerging and becoming green.

cruciate: cross-shaped.

cyclic: arranged in whorls (rings).

cymose: of or having inflorescence with all branches of limited growth; *n.* cyme.

deciduous: dropping off.

dehiscence: the shedding of seeds or spores.

dehiscent: of or having fruit which stays on plant and opens to release seeds.

dendroid: freely branched, tall with erect main stem.

dichotomous: of or having branching where an apical cell divides into 2 equal parts and each gives rise to a branch; of or having similar venation.

dictyostele: a stele in which the leaf gaps are close together; *a.* dictyostelic.

differentiation: the formation of tissues.

diffuse: of or having growth where all cells are capable of division.

dikaryotic: of or having cells of hypha containing 2 nuclei of different mating types.

dioecious: bearing male and female sex organs on different plants.

diplanetic: producing 2 types of zoospore.

diploid: having 2 sets of chromosomes; symbol 2n.

diploxylic: having an internal and external strand of xylem.

disc: the fleshy, sometimes nectar-secreting portion of the receptacle.

dolipore: of or having septa which are inflated and centrally perforated as in basidiomycetes.

double fertilization: in angiosperms, the condition of having 2 male gametes in the pollen tube, one of which fuses with the female gamete to form a zygote, the other of which fuses with the secondary nucleus to form a triple fusion nucleus from which the endosperm develops.

drupe: a (usually) fleshy fruit with usually one seed which is surrounded by a stony layer that is part of the fruit wall.

elaters: in fungi, filaments of the capillitium; in liverworts, spirally thickened cells which are mixed with spores and are important in spore dispersal; similar structures are found in some horsetails.

endogenous: formed inside an organ.

endoparasite: internal parasite.

endophyte: a plant living within another plant; *a*. endophytic.

endosperm: the nutritive tissue in seed of angiosperms and gymnosperms, in angiosperms made from the triple fusion nucleus and in gymnosperms from the female prothallus.

endospore: a type of spore rather like an aplanospore.

endosporic: having prothallus retained within the spore wall and scarcely protruding from it, even at maturity when the spore breaks open.

epicalyx: a calyx-like structure outside but close to the true calyx.

epigynous: having an inferior ovary.

epipetalous: attached to the petals.

epiphyte: a plant growing on the surface of another plant; *a*. epiphytic.

eucarpic: having only part of the thallus forming a fruit body.

euglenoid movement: wriggling movements produced by contractions of the whole cytoplasm within the pellicle.

eusporangiate: having sporangia with sporogenous tissue derived from inner daughter cell.

evanescent: soon disappearing.

exospore: a spore borne externally; in blue-green algae, a modified endospore.

exosporic: having prothallus breaking through spore wall and developing outside the spore.

exstipulate: without stipules.

eyespot: a light-sensitive pigmented region in unicellular algae; also called stigma.

facultative: occasional.

fascicle: cluster or bundle.

ferruginous: rust-coloured.

filament: in flowering plants, the stalk of a stamen; in blue-green algae, a trichome surrounded by a mucilaginous sheath; in general, a line of cells joined together; *a*. filamentous.

fission: splitting into 2 or more parts.

flagellum: a whip-like protoplasmic extension of cytoplasm by which some cells swim by lashing movements; *plu*. flagella.

foliaceous: leaf-like.

follicle: a dry, dehiscent fruit formed from a single carpel, of which there are several in a flower, dehiscing along one side, and usually containing several seeds.

fragmentation: reproduction by breaking off of a part, especially of a filament.

free-central placentation: the arrangement of ovule so that it stands up from the base of the ovary and is not united with walls or top.

frond: the leaf of a fern; the flattened part of the thallus of a seaweed.

fruit: in angiosperms, a structure made from the ovary after fertilization, and sometimes also from associated structures, and containing the seeds.

fruit body: a structure bearing reproductive bodies, especially spores; also called fructification, sporocarp.

frustule: cell wall of a diatom, made of 2 overlapping halves like box and lid, the larger part (lid) called epitheca, and smaller part called hypotheca; also used to include the diatom's living contents.

gametangium: a gamete-producing structure; *plu.* gametangia.

gamete: one of 2 haploid cells which fuse in sexual reproduction to form a zygote; also called sex cell.

gametophyte: the haploid generation in a plant life cycle, which bears the gametes.

gas vacuole: a vacuole believed to contain gas.

gleba: interior of the basidiocarp in Gastromycetes.

Gram negative: of bacteria which, when stained with a basic dye then washed with alcohol or acetone, are decolorized.

Gram positive: of bacteria which, when stained with a basic dye then washed with alcohol or acetone, retain the colour.

gynoecium: the female part of the flower, made of one or more carpels, fused or free, making one or more stigmas, styles and ovaries; also spelt gynaecium.

haplocheilic: having stomata in which the guard cells and subsidiary cells are not derived from the same mother cell.

haploid: having one set of chromosomes; symbol n.

haptonema: an anchorage organ in some Haptophyceae.

haustorium: a hypha secreting enzymes and used for feeding; *plu.* haustoria.

herbaceous: soft, green, non-woody.

hermaphrodite: having male and female sex organs in the same flower, or combined in the same reproductive structure.

heterocyst: in blue-green algae, an enlarged vegetative cell whose function is unknown, but which may be concerned with formation of hormogonia.

heteromorphic: of or having alternation of generations in which the sporophyte and gametophyte are different in appearance.

heterosporous: having spores of 2 distinct sizes; *n.* heterospory.

heterotrichous: having an erect and prostrate system.

heterotrophic: obtaining food from complex organic materials and including holozoic, parasitic and saprophytic nutrition.

holdfast: an organ attaching a plant, especially an alga, to a substratum.

holocarpic: having all of the thallus involved in the formation of the fruit body.

holozoic: feeding like an animal, by engulfing solid food particles and digesting them internally.

homosporous: having all spores of approximately the same size; *n.* homospory.

hormogonia: the portions into which a filamentous blue-green alga breaks up during reproduction; *sing.* hormogonium; also called hormogones.

hyaline: translucent but still characteristically coloured.

hygroscopic: taking up or losing water, resulting in movement.

hymenium: the layer in a fruit body made up of the actual cells (asci or basidia) which produce the spores (ascospores or basidiospores).

hypha: a filament of the plant body (mycelium) of a fungus; *plu.* hyphae; *a.* hyphal.

hypogynous: having a superior ovary.

hypothallus: a film-like residue that remains after the formation of sporangia in slime moulds.

indehiscent: of or having fruit which does not open to release seeds, so whole fruit falls off the plant.

indusium: a covering protecting the sorus of ferns; *plu.* indusia.

inferior: of or having ovary with perianth inserted around the top, the ovary being sunk and fused to receptacle.

inflorescence: all the flowers on one branch or stem; in gymnosperms, complex strobili surrounded by perianth.

inoperculate: without a lid or terminal pore.

integument: a coat of an ovule.

intercalary: in a position other than at the tip.

internode: the length of stem between 2 nodes.

involucre: a cluster of bracts forming a calyx-like structure.

isogamy: the fusion of gametes of similar size and form; *a.* isogamous.

isomorphic: of or having alternation of generations in which the sporophyte and gametophyte are similar in appearance.

lamina: a blade, as of a leaf or seaweed frond.

latex: milky juice.

legume: a fruit in Leguminosae, consisting of a single carpel, containing several seeds and opening by splitting along the midrib.

leptosporangiate: having sporangia with sporogenous tissue derived from outer daughter cell.

ligule: a small scale borne on the surface of a leaf or perianth segment.

locule: a compartment or cavity; also called loculus.

macrozoospore: large zoospore.

manoxylic: of or having wood soft, sparse and loose with wide parenchyma rays.

mazaedium: a fruit body consisting of a powdery mass of free spores, interspersed with sterile threads, enclosed in a walled structure.

medulla: the central zone in stipe or frond of a seaweed.

megaphyllous: of or having large leaves; also called macrophyllous.

megasporangium: a sporangium in which meiosis occurs to produce megaspores; in spermatophytes corresponds to nucellus of ovule; *plu.* megasporangia.

megaspore: a spore produced by meiosis, formed in some pteridophytes and in spermatophytes, containing the female gametophyte generation; in spermatophytes also called embryo sac.

megasporophyll: a sporophyll bearing megasporangia; in gymnosperms sometimes called an ovuliferous scale, and in angiosperms called a carpel.

megastrobilus: a cone-like structure made of megasporophylls; also called female cone; *plu.* megastrobili.

meiosis: nuclear or cell division resulting in halving the chromosome number and producing 4 haploid, non-identical cells.

meiospores: spores produced by meiosis.

microphyllous: of or having small leaves.

micropyle: the hole in the integument and seed coat.

microsporangium: a sporangium in which meiosis occurs to produce microspores; in spermatophytes also called pollen sac; *plu.* microsporangia.

microsporangiophore: a non-leaf-like structure bearing microsporangia.

microspore: a spore produced by meiosis which contains the male gametophyte generation; in spermatophytes also called pollen grain.

microsporophyll: a sporophyll bearing microsporangia, in angiosperms modified to form a stamen.

microstrobilus: a cone-like structure made of microsporophylls; also called male cone; *plu.* microstrobili.

microzoospore: small zoospore.

mitosis: nuclear or cell division in which the chromosome number remains the same, and which produces 2 identical cells.

monocolpate: of or having pollen grains with a single elongate aperture (colpus).

monoecious: bearing male and female sex organs, or male and female flowers on the same plant, but not in the same reproductive structure.

monoporate: with one pore.

monospore: a spore formed instead of a tetraspore, and with no meiosis.

monosporous: derived from one spore.

multiaxial: a type of construction in red algae, having several main filaments running parallel longitudinally.

multilocular: having many compartments.

multiseriate: arranged in several rows.

mycelium: the mass of hyphae making up the plant body in fungi.

myxamoebae: motile amoeba-like cells; *sing.* myxamoeba.

myxoflagellatae: motile flagellated cells; *sing.* myxoflagellata.

n: haploid; **2n:** diploid.

nannocytes: motile, small, naked protoplasts formed by division of cell contents without any visible enlargement of vegetative cell.

nectary: a structure from which nectar is secreted.

node: a point on a stem where one or more leaves arise.

non-sexual: of or having reproduction which does not involve the fusion of gametes; also called asexual.

nucellus: the tissue in the ovule in which meiosis occurs.

nut: a dry, indehiscent fruit with a hard woody fruit wall, formed from a syncarpous ovary and usually containing a single seed.

nutlet: a one-seeded portion of a fruit which breaks off as it matures.

obligate: invariably.

oidium: a thin-walled free hyphal cell behaving as a spore; *plu.* oidea.

oogamy: the fusion of a large non-motile female gamete and a small motile male gamete; *a.* oogamous.

oogonium: the female sex organ in algae and fungi; *plu.* oogonia.

oospore: a thick-walled resting spore formed after the fusion of differentiated gametangia; a zygote formed by oogamy.

operculate: with a lid.

operculum: a lid or lid-like structure.

ostiole: a pore, especially in a reproductive stucture.

ovary: the part of the gynoecium enclosing the ovules.

ovule: in spermatophytes, a structure containing the ovum, which after fertilization develops into a seed.

ovuliferous: bearing ovules.

ovum: a female gamete.

palmate: consisting of more than 3 leaflets arising from the same point.

palmelloid: of or having a colonial stage in which cell division occurs within a cyst, with the organism in a resting stage.

papillose: having small elongated projections (papillae).

paracarpous: having fused carpels but no (or incomplete) septa, and parietal or free-central placentation.

paraphyses: thread-like structures often associated with reproductive organs or cells in lower plants; *sing.* paraphysis.

parasitic: deriving food from another living organism (host) to which it is attached, and harming the host.

parietal: at the edge or periphery.

parietal placentation: position of ovules in ovary where ovules arise from placentae on peripheral wall of the carpels.

pellicle: a thin outer covering, not a cell wall.

peltate: applied to a flat organ with stalk inserted on under-surface, not at edge, like a nasturtium leaf.

pendulous: hanging downwards.

peptidoglycan: a heteropolymer unique to cell wall of prokaryotes, consisting of 2 amino sugars, N-acetyl-glucosamine and N-acetyl-muramic acid, and a peptide made of a few amino acids; also called mucopeptide, murein.

perianth: in angiosperms, floral leaves as a whole, petals and sepals; in gymnosperms, bract-like structures surrounding micro- and megasporophylls; in bryophytes, a tube-like or purse-like structure around archegonium, especially in liverworts.

peridiole: one of the discrete portions, 'eggs', into which the gleba of a bird's nest fungus is broken up.

peridium: in Gastromycetes, the membrane, usually of at least 2 layers, that surrounds the fruit body; in slime moulds, the outer calcified coat of the fruit body.

perigynous: of or having flowers intermediate between hypogynous and epigynous.

periplasm: a peripheral layer of protoplasm.

periplast: a thin outer covering, not a cell wall.

perisperm: nutritive tissue derived from the nucellus of a seed.

peristome: the ring of teeth that surrounds the mouth of a moss capsule, and is visible when the lip (operculum) of the capsule is removed.

perithecium: a flask-shaped ascocarp with a pore in the top.

petal: a member of the inner series of perianth segments, if differing from the outer, especially if brightly coloured.

petaloid: petal-like.

petiole: leaf-stalk.

phloem: tissue, chiefly conducting sugars, in vascular plants, consisting of sieve tubes and (in angiosperms only) companion cells.

phyllids: leaf-like structures in bryophytes, not true leaves because they do not have a cuticle, stomata or vascular tissue.

physiological anisogamy: the fusion of gametes differing in behaviour but not in form.

pinna: a leaflet of a pinnate leaf; *plu.* pinnae.

pinnate: a leaf composed of more than 3 leaflets arranged in 2 rows along a common stalk.

pinnule: a lobe or segment of a pinna or of a pinnate leaf.

pit: a hole in a cell wall.

placentation: the position of placentae (the places at which ovules arise) in an ovary.

plasmodium: a jelly-like mass of aggregated cells, in slime moulds.

pleurocarpous: bearing archegonia, and hence the sporophyte generation, on a short side branch, not at the tip of the main stem, and having a prostrate habit.

plumule: embryonic shoot.

plurilocular: having many compartments.

pollen grain: a microspore in a seed plant.

polycyclic stele: a stele in which 2 or more steles occur within each other.

polygamous: having male, female and hermaphrodite flowers on the same or different plants.

polyporate: having many pores.

polysiphoneous: of or having a filament in which the main central filament is surrounded by peripheral cells.

polystelic: having more than one stele visible in any cross section.

primary growth: growth from the apical dividing cells (meristems), not from the cambium, and producing soft herbaceous tissue.

promycelium: a short hypha produced by a fungal spore, on which a different kind of spore develops.

propagule: a vegetative structure which becomes detached from parent and grows into a new plant.

prostrate: creeping.

prothallus: the gametophyte generation in pteridophytes and gymnosperms; *plu.* prothalli.

protonema: a filamentous stage produced from a spore, before the development of a thalloid body.

protoplast: a unit consisting of a nucleus, cytoplasm and cell membrane, with or without a cell wall.

protostele: a stele without leaf gaps; *a.* protostelic.

pseudocapillitium: a structure similar to the capillitium, but of different origin.

pseudofilamentous: having colonial cells giving the appearance of filaments.

pseudoparaphyses: sterile hyphae with degenerate nuclei on a hymenium; *sing.* pseudoparaphysis.

pseudoparenchyma: a mass of closely interwoven hyphae or filaments, appearing like parenchyma in section; *a.* pseudoparenchymatous.

pseudoplasmodium: a mass of closely associated myxamoebae which have not united to form a true plasmodium.

pseudopodium: in some mosses, a leafless prolongation of the stalk of a gametophyte on which the capsule is borne, instead of on a seta; *plu.* pseudopodia.

pseudoraphe: a lighter region on the surface, giving a superficial appearance of a raphe.

pseudothecium: ascostroma, especially in lichens; *plu.* pseudothecia.

pulvinate: having enlarged and swollen leaf bases (pulvini).

pusule: a small contractile vacuole, or a non-contractile vacuole which discharges to exterior by a duct.

pycnidiospore: a spore formed within a pycnidium.

pycnidium: a roundish fruit body, rather like a perithecium, but not involving sexual reproduction; *plu.* pycnidia.

pycnoxylic: of or having wood relatively dense, with small parenchyma rays.

pyrene: a small hard body containing a single seed, like a drupe stone, but having several together in one fruit.

pyrenoid: a small protein body in a chloroplast, around which starch is formed.

raceme: a more-or-less conical, unbranched inflorescence in which the oldest flowers are at the bottom and the youngest at the top.

racemose: like a raceme.

radial symmetry: having many planes of symmetry, i.e. forming mirror image halves if cut along any diameter; in flowers also called actinomorphic.

radicle: embryonic root.

raphe: a slit running along the longitudinal axis from one polar nodule to the other in the frustule of some pennate diatoms.

receptacle: in some algae, a structure in which conceptacles bearing sex organs are housed; in some gymnosperms, the structure on which ovules and later seeds are borne; in angiosperms, the upper part of the flower stalk from which the floral parts arise.

reticulate: like a net.

rhizoid: a unicellular or filamentous structure acting as a root-like organ; *a.* rhizoidal.

rhizome: a horizontal, usually underground, stem; *a.* rhizomatous.

rhizophore: a structure running from the stem and ending in a root, in some club mosses.

rhizopodal: having the form of an amoeba's pseudopodium.

root cap: a hollow cone of cells covering the growing tip of a root and protecting it from damage as it pushes through the soil.

saccate: pouched.

samara: a dry, indehiscent fruit, part of the wall of which forms a flattened wing.

saprophytic: deriving food from dead or organic material.

schizocarp: a fruit which splits into separate one-seeded portions when mature.

sclerotium: a hard resting body resistant to unfavourable environmental conditions.

secondary growth or thickening: growth produced from the cambium and cork cambium, and including wood and bark.

seed: in spermatophytes, a reproductive structure consisting of an embryo, a food store (endosperm) and a coat, and developing from an ovule.

sepal: a member of the outer series of perianth segments, especially when green and leaf-like.

sepaloid: sepal-like.

septate: having partitions (septa).

sessile: not stalked.

seta: the diploid stalk, part of the sporophyte, on which the capsule is borne in some bryophytes.

sex organ: a structure in which gametes are produced; sometimes called gametangium.

sexual: of or having reproduction involving fusion of gametes to form a zygote.

silicification: impregnation with silica; *a.* silicified.

silicula: a capsule-like fruit made of 2 carpels, almost as wide as it is long, in Cruciferae.

siliqua: a long thin capsule-like fruit made of 2 carpels, in Cruciferae.

solenostele: a stele in which the leaf gaps are far apart; *a.* solenostelic.

somatic: not reproductive.

sorus: a cluster of sporangia; *plu.* sori.

spadix: a spike with a swollen fleshy axis enclosed in a spathe.

spathe: a large, usually coloured bract, which more-or-less encloses a spadix.

sperm: spermatozoid.

spermatium: a non-motile male gamete, in red algae; *plu.* spermatia.

spermatozoid: a motile male gamete in many lower plants; abbreviated to sperm; also called antherozoid.

spike: a simple racemose inflorescence with sessile flowers.

spirocyclic: arranged partly in spirals and partly in whorls (rings).

sporangiolum: a small deciduous sporangium containing one or very few spores; *plu.* sporangiola.

sporangiophore: a hypha or filament bearing a sporangium.

sporangiospore: a spore borne within a sporangium.

sporangium: a structure containing spores; *plu.* sporangia.

spore: a small unicellular or few-celled dispersive and usually reproductive body.

sporiferous: spore-producing or spore-bearing.

sporocarp: a multicellular structure in or on which spores are formed; also called fruit body or fructification.

sporodochium: a structure made of conidia with spore mass supported on a bed of short conidiophores.

sporogenous: spore-producing or spore-bearing.

sporophore: a spore-bearing structure, either organized as a sporocarp or a simpler structure.

sporophyll: a leaf-like structure, or one regarded as homologous with a leaf, bearing or subtending sporangia.

sporophyte: the whole of the spore-bearing non-sexual diploid generation in a life cycle with alternation of generations.

sporulating: producing spores.

stamen: in angiosperms, a structure evolved from a microsporophyll and composed of an anther (containing pollen sacs) and filament (stalk).

staminode: a sterile, often reduced, stamen.

statospore: a thick-walled cyst formed from contents of cell within mother cell.

stele: the primary vascular tissue in vascular plants, consisting of xylem, phloem, pericycle and endodermis.

stellate: star-shaped.

stigma: the receptive surface on a gynoecium to which the pollen grains adhere; in algae, another name for eyespot.

stipe: stalk of the fruit body of a fungus or thallus of a seaweed.

stipulate: having stipules.

stipule: a scale-like or leaf-like appendage, usually at the base of the petiole and sometime attached to it.

stoma: a pore in the epidermis which can be closed by changes in shape of surrounding guard cells; *plu.* stomata.

striae: long narrow depressions or ridges; *a.* striate.

strobilus: a group of sporophylls with their sporangia, more-or-less tightly packed around a central elongated axis, forming a well defined group; also called cone; *plu.* strobili.

stroma: a compact structure of hyphae on or in which fruit bodies are formed; *plu.* stromata.

style: the part of the gynoecium connecting the ovary with the stigma.

subsidiary cell: a cell associated with the guard cell of a stoma, but differing in structure from both it and from normal epidermal cells.

substratum: non-living material to which a plant is attached; also called substrate.

superior: of or having an ovary with perianth inserted around the base.

symbiosis: an internal partnership between 2 organisms (symbionts) for their mutual benefit; *a.* symbiotic.

sympetalous: with fused petals.

synangium: a group of fused sporangia; *plu.* synangia.

synascus: a compound ascus.

syncarpous: broadly, having carpels united to one another; more strictly, having fused carpels separated by septa and with axile placentation.

syndetocheilic: having stomata in which both subsidiary cells and guard cells are formed from a single mother cell.

synnema: a group of erect, sometimes fused conidiophores bearing conidia; *plu.* synnemata.

tap root: the main root developed from the radicle of the seed.

teliospore: a thick-walled resting spore capable of quiescence and giving rise to promycelia on which basidiospores develop; also called teleutospore.

tetrasporangium: a sporangium containing tetraspores.

tetraspore: a non-sexual spore in red algae, produced in groups of 4, usually but not always by meiosis; in brown algae, a large non-motile spore produced in a unilocular sporangium.

tetrasporophyte: in red algae, the generation in the life cycle producing tetraspores.

thallus: a multicellular body not differentiated into stem, root and leaves; *a.* thalloid.

thylakoid: one of the membranes on which chlorophyll is situated, usually in chloroplasts.

tinsel flagellum: a flagellum with short side branches along the central axis; also called pantonematic, flimmer.

tracheid: a single elongated water-conducting cell with pointed ends and no living contents.

trichocyst: a minute hair-like body exuded from cells in some algae.

trichogyne: the elongated terminal portion of the female sex organ in some lower plants, which receives the male gamete or sex organ.

trichome: in Schizophyta, a filament of cells, not surrounded by a mucilaginous sheath.

trichospore: a flagellated or ciliated spore.

trichothallic: of growth in algae by a few cells towards or at the base of the filament.

tricolpate: of or having pollen grains with 3 elongated apertures (colpi).

trilocular: with 3 compartments.

umbel: an inflorescence in which the flower stalks arise from the same point at the top of the stem, forming an umbrella-shaped inflorescence.

uniaxial: a type of construction in some algae having a single main filament and growing point.

unilocular: having a single compartment.

uniseriate: arranged in a single row.

unisexual: having organs of one sex only.

unitunicate: having wall of one layer.

uredospore: an orange or brownish spore formed for rapid propagation.

vascular bundle: xylem and phloem found together in a strand, sometimes separated by cambium.

vascular plant: a plant with vascular tissue.

vascular tissue: conducting tissue of xylem and phloem.

vegetative reproduction: non-sexual reproduction by multicellular propagules rather than spores.

vegetative stage: non-reproductive stage.

venation: arrangement of veins in a leaf.

vesicle: a thin-walled globular swelling, usually at end of hyphae.

vessel: a long water-conducting tube of xylem in which cells are thickened and dead, and cross walls have broken down to form a continuous tube.

volva: a sheath of hyphae enclosing the whole fruit body in some agarics, which becomes ruptured as fruit body enlarges.

whiplash flagellum: a smooth flagellum; also called acronematic.

whorl: more than 2 organs of the same kind, at the same level, forming a ring.

wood: tissue composed of xylem, particularly secondary xylem.

woody: having secondary xylem.

xeromorphic: adapted to dry conditions.

xylem: water-conducting tissue in vascular plants, made of vessels and tracheids.

zonate: of or having tetraspores in a row of 4, not a tetrahedral group.

zoosporangium: a structure producing zoospores; *plu.* zoosporangia.

zoospore: a non-sexual, motile reproductive cell, which swims by one or more flagella; also called swarm cell, swarmer.

zygomorphic: bilaterally symmetrical.

zygospore: a thick-walled resting spore formed by fusion of undifferentiated gametangia, isogamy or anisogamy.

zygote: a diploid cell formed by fusion of 2 gametes in sexual reproduction.

References and further reading

AHMADJIAN, V. and HALE, M. E., *The Lichens*. Academic Press, 1973.

AINSWORTH, G. C., *Ainsworth and Bisby's Dictionary of the Fungi*, 6th edn including 'The Lichens' by James, P. W. and Hawksworth, D. L. Commonwealth Mycological Institute, 1971.

ANDREWS, H. N., *Studies in Palaeobotany*. John Wiley, 1961.

ARNOLD, C. A., *An introduction to Palaeobotany*. McGraw-Hill, 1947.

BERGEY, D. H., *Bergey's Manual of Determinative Bacteriology*, 7th edn by Breed, R. S., Murray, E. G. D. and Smith, N. R. Baillière, N. R. Tindall and Cox, 1957.

BERGEY, D. H., *Bergey's Manual of Determinative Bacteriology*, 8th edn by Buchanan, R. E. and Gibbons, N. E. Williams and Wilkins, 1974.

BOLD, H. C., *The Plant Kingdom*, 3rd edn. Prentice-Hall, 1970.

CHAPMAN, V. J. and D. J., *The Algae*, 2nd edn. Macmillan, 1973.

CLAPHAM, A. R., TUTIN, T. G. and WARBURG, E. F., *Flora of the British Isles*, 2nd edn. Cambridge University Press, 1962.

CLARKE, G. C. S. and DUCKETT, J. G. *Bryophyte Systematics*. Academic Press, 1979.

CRONQUIST, A., *The Evolution and Classification of Flowering Plants*. Thomas Nelson, 1968.

DAVIS, P. H. and HEYWOOD, V. H., *Principles of Angiosperm Taxonomy*. Oliver and Boyd, 1963.

HALE, M. E., *The Biology of Lichens*, 2nd edn. Edward Arnold, 1974.

INGOLD, C. T., *The Biology of Fungi*, 2nd end. Hutchinson, 1969.

International Code of Botanical Nomenclature: adopted by the 12th International Botanical Congress, Leningrad, July 1975. 1978.

JEFFREY, C., *Biological Nomenclature*, 2nd edn. Edward Arnold, 1977.

JONES, S. B. and LUCHSINGER, A. G., *Plant Systematics*. McGraw-Hill, 1979.

LEEDALE, G. F., *Euglenoid Flagellates*. Prentice-Hall, 1967.

MACVICAR, S. M., *Student's Handbook of British Hepatics*. Wheldon and Wesley, 1926, reprinted 1955.

MORRIS, I., An *Introduction to the Algae*, 2nd edn. Hutchinson, 1971.

PARIHAR, N. S., *An Introduction to Embryophyta, Vol. I. Bryophyta*, 5th edn. Allahabad: Central Book Depot, 1965.

PRESCOTT, G. W., *The Algae: A Review*. Thomas Nelson, 1969.

ROUND, F. E., *The Biology of the Algae*, 2nd edn. Edward Arnold, 1973.

ROUND, F. E., *Introduction to the Lower Plants*. Butterworths, 1969.

SCHUSTER, R. M., *The Hepaticae and Anthocerotae of North America*. Columbia University Press, 1966.

SCOTT, G. A. M., STONE, I. G. and ROSSER, C., *The Mosses of Southern Australia*. Academic Press, 1976.

SMITH, A. J. E. *The Moss Flora of Britain and Ireland*. Cambridge University Press, 1978.

SMITH, G. M., *Cryptogamic Botany, Volumes I and II*, 2nd edn. McGraw-Hill, 1955.

SPORNE, K. R., *The Morphology of Gymnosperms*. Hutchinson, 1965.

SPORNE, K. R., *The Morphology of Pteridophytes*, 2nd edn. Hutchinson, 1966.

SPORNE, K. R., *The Mysterious Origin of Flowering Plants*. Oxford Biology Readers: Oxford University Press, 1971.

STEARN, W. T., *Botanical Latin*. Thomas Nelson, 1966.

STRASBURGER, E., *Strasburger's Textbook of Botany*, trans. from the 30th German edn by Bell, P. and Coombe, D. Longman, 1976.

TAKHTAJAN, A., *Flowering Plants – Origin and Dispersal*, trans. by Jeffrey, C. Oliver and Boyd, 1969.

WATSON, E. V. *The Structure and Life of Bryophytes*, 3rd edn. Hutchinson, 1971.

WATSON, E. V., *British Mosses and Liverworts*, 2nd edn. Cambridge University Press, 1968.

WILLIS, J. C., *A Dictionary of the Flowering Plants and Ferns*, 6th edn. Cambridge University Press, 1951.

WILLIS, J. C., *A Dictionary of the Flowering Plants and Ferns*, 8th edn revised by Airy Shaw, H. K. Cambridge University Press, 1973.

Index